Investigating Biological Concepts

A LABORATORY MANUAL

By John Dickerman
Northern Illinois University

Morton
Publishing Company

925 W. Kenyon, Unit 12
Englewood, Colorado 80110

Book Team

Publisher	Douglas Morton
Executive Editor	Chris Rogers
Copy Editor	Kevin Campbell, Greenleaf Editorial
Production/Design	Focus Design

Credits

Figure 6.2, 6.3 and 12.3 Courtesy of Patrick Guilfoile, *A Photographic Atlas for the Molecular Biology Laboratory*, Morton Publishing Company, courtesy Audrey Guilfoile.

Figure 3.1 Courtesy of Leica Microsystems, Inc., Deerfield, IL

Figure 20.4 and 20.15 Courtesy of Champion International Corporation

Figure 18.3, 18.4, and 19.6 from Kent M. Van De Graaff & John L. Crawley, *A Photographic Atlas for the Biology Laboratory*, Morton Publishing Company, courtesy of John L. Crawley.

Figure 20.6 from Kent M. Van De Graaff & John L. Crawley, *A Photographic Atlas for the Botany Laboratory*, Morton Publishing Company, courtesy of John L. Crawley.

Figure 21.8, 21.9, and 21.12 from Kent M. Van De Graaff & John L. Crawley, *A Photographic Atlas for the Zoology Laboratory*, Morton Publishing Company, courtesy of John L. Crawley.

Printed in the United States of America
by Morton Publishing Company
925 W. Kenyon Ave., Unit 12, Englewood, CO 80110

10 9 8 7 6 5 4 3 2 1

ISBN: 0-89582-508-2

Acknowledgements

Writing a textbook covering the full range of general biology – even at an introductory level – is a daunting task. There is no doubt that the topics included in this book range far beyond the areas I can claim any kind of expertise in. Luckily, while on the faculty at Northern Illinois University, I have been able to work with many talented scientists who have the expertise I lack. I owe a special debt to Professor Arthur Weis (currently at the University of California) who taught me a great deal both about biology and how to teach it. His influence can be found throughout this book, but especially in chapters 13–15, which are based on exercises he taught me years ago. I have also benefited greatly from working with Professor Ronald Toth, who showed me how to make the study of diversity into a fascinating story by organizing it upon a historical, evolutionary framework. Over the years I have also garnered many useful ideas from discussions with Professors Jack Bennett, Mitrick Johns, Bethia King, and Steve Nadler (who has also moved on to the University of California). Patrick McCarthy has always been indispensable when I had questions about microbiology, and Daniel Olson has given me much helpful advice and support. It is impossible to list all of the graduate teaching assistants who have helped me to get these exercises working, but I need to make special mention of Mary Crowe, James Robins, Mike Palm, and Ted Chauvin for their especially useful advice. I won't even attempt to name the scores of people on the Biolab listserv with whom I have corresponded over the years, but to all of them I also owe a great deal of thanks.

This book would not have happened had it not been for my editor at Morton, Chris Rogers, who has been wonderful to work with. And it has been much improved by the many thoughtful comments of Professor Merle Heidemann of Michigan State University, who took the time to review the manuscript.

I must further acknowledge sources of information and inspiration for the following chapters:

The manometer described in chapter 8 is a scaled-down version of that described by L. James McElroy in the March 1976 issue of *The American Biology Teacher*.

The recipe for the BZ reaction in chapter 8 is based mainly on that published by Arthur T. Winfree in *The Geometry of Biological Time* (New York: Springer-Verlag, 1980).

Values for the length of onion root mitotic stages in chapter 10 are those published by Marshall D. Sundberg (citing N. S. Cohn's 1969 *Elements of Cytology*) in the October 1981 issue of *The American Biology Teacher*.

To all of these people, I am grateful. Of course, I remain completely responsible for the text, and any errors or inaccuracies are wholly my own.

Preface to the Students

LEARNING HOW TO *DO* BIOLOGY

This book is intended to complement the (undoubtedly large) textbook you use for the lecture portion of your class. You will find that it covers similar material, but is focused on hands-on exercises. The sometimes lengthy introductions are offered to help you bridge the gap between lecture and lab, so that you can actively apply what you have learned to your laboratory work. A fundamental principle of modern science is that a true scientist should take nothing simply on faith, but should demand evidence of any assertion. To the extent that it is possible in an introductory course, this book is designed to allow you to directly experience first-hand many of the phenomena that you read about.

As a result of this approach, you will find the exercises in this book to be a varied lot indeed. Some will be simple and fool-proof, while others will be complex and challenging. Some phenomena will come out in dramatic, even colorful, displays; others will only reveal themselves after numerous measurements and sophisticated calculations. Although written for the beginning student, these experiments are intended to mirror the practice of biology in modern research laboratories. As you work through them you should get a better idea of which areas of biology you are suited for and which areas interest you least.

This book is built around several central *concepts* of biology. Studying biology means more than simply memorizing lists of facts. As you will see in chapter 1, science is a logical process by which we make generalizations about our observations. This means that you will have to *think* about what you see in the lab. It is very important that you realize that an exercise does not end when you leave the lab – you must continue to reflect upon your results and question what happened if you are to really *understand* what you experienced.

HOW TO STUDY

At the beginning of each chapter are three sets of objectives to let you know what is expected of you. (Chapter 1 has only two sets, as it introduces the processes necessary for the conceptual objectives of subsequent chapters.) The first set describes what *new skills* you will be learning in the exercises. These are the basic scientific techniques – using a microscope, testing solutions, sampling populations, etc. – that a beginning biology student must learn. You can only learn these skills by doing them, and you will only improve them with practice. Likewise, your proficiency in these skills can only be assessed by some kind of practical method: your instructor may ask you to demonstrate a technique or require you to identify some unknown specimen by using the technique. There is no realistic way to assess these skills through a written test.

The second set of objectives describes the basic facts you must learn for each topic. In most cases this means learning the terms (and their meanings) listed at the end of each chapter. These terms are given at least a brief definition in each chapter, but it will often be helpful for you to compare these definitions with those in your lecture notes or text in order to get a better idea of how the term is used. Learning terminology is essentially rote learning. It is facilitated by repetitive drill and practice. This is not the most exciting part of the exercises, but it is important for building up a basic vocabulary so you can discuss what you see. One way to make this type of learning more interesting is to work with a study partner so you can drill each other. This type of knowledge is easily assessed by written quizzes and tests. What you learn in the lab should carry over into the lecture portion of your class.

The final set of objectives is generally considered the most important. These objectives are also the hardest to achieve. They are stated in terms of what you will observe in a given exercise, but they go far beyond observation. These concepts reflect the generalizations and applications you will make from your observations. There is no simple way to learn concepts. You must make observations and then compare them to tease out the underlying principles. As you perform an exercise, try to generalize what you see. Make predictions for other situations. Most exercises include questions to help you with this process. Conceptual learning can be assessed in various ways, but such assessment will always involve asking you to explain something or predict what will happen in a similar situation. Your instructor may ask you to keep a notebook of your work (see appendix 1). The "Conclusions" section of each notebook entry is a place for you to write your explanations of your observations. Short essays or discussion sessions may be used as forums for presenting conclusions.

Table of Contents

The Logical Basis of Science

1

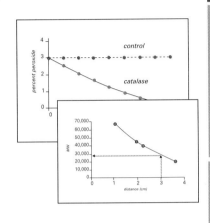

OBJECTIVES

New Skills

After working on the exercises in this chapter you will begin to be able to
1. make logical generalizations from a set of data.
2. apply generalizations (theories) to specific situations to make predictions (deductions or hypotheses).
3. suggest appropriate controls when planning an experiment to test a hypothesis.
4. express numerical data in graphic form and use these graphs to extrapolate predictions for new conditions.

Factual Content

In this chapter you will learn to use terminology dealing with logic and the scientific process.

INTRODUCTION

The word *science* is derived from a Latin term for knowledge. Science is a special kind of knowledge, however. For one thing, it is **empirical**, which means that it is based on observation and experience. (This separates science from other branches of knowledge—for example, religion—which are free to deal with things that cannot be observed.) A "scientific fact" is any observation that is **reproducible** – that is, an observation that can be repeated and verified by other observers under specified conditions.

But science is more than just a catalog of facts, for scientists spend a great deal of effort organizing the facts they discover within a logical framework. **Logic** is the study of how to build consistent and orderly arguments from a set of elementary facts or propositions. In practice, logic allows us to make and apply valid generalizations based on the observations we make. Science, therefore, includes not only the collection and verification of observations, but also the process of making generalizations from those observations and using these generalizations to predict what might happen in a new situation that no one has observed before.

LOGICAL REASONING

Inductive logic refers to the process of making generalizations from basic facts or principles. All of the grand theories of science are products of inductive reasoning: The theory of gravity is built upon observations of falling objects and planetary orbits; atomic theory is based on measurements of gas behavior and studies of chemical reactions; and the theory of evolution is derived from observations of species diversity, shared characteristics between species, and fossils. Logical induction is central to science, since no theory can be considered scientifically valid unless it has a strong empirical basis of observable fact. Unfortunately, inductive reasoning has a weakness: no inductive theory can be "proven" in any absolute sense unless every possible case is observed. For example, a person who has only seen a few birds, say crows, robins, and blackbirds, might generalize:

All birds have black feathers.

This statement is true for all of the observed cases, but is it *always* true? What if the next bird this person saw was a white dove? An inductive theory cannot be absolutely proven, but it may be disproved by a single counter-example. Scientists must always be open to revising their theories as new evidence accumulates.

Deduction is the opposite of induction – starting with a generalization, one may deduce, or predict, what would then happen in a specific case. The fictional detective Sherlock Holmes was famous for using his vast knowledge of criminology to deduce the solutions of specific mysteries. For example, in one story, he used the generalization

The dog always barks at strangers.

to deduce that

The dog would have barked had the crime been committed by a stranger.

Being specific, a deduction can be tested by observation. In the Sherlock Holmes story, witnesses told the detective that the dog had not barked, leading the detective to conclude that the crime had, in fact, been committed by someone that the dog knew.

Scientists use the methods of induction and deduction together: Initial observations lead to a generalization. That generalization can then be tested by using it as the basis for specific deductions and then performing further observations to determine whether the deductions were accurate predictions. A scientific prediction is often called a **hypothesis**, and an investigation designed to test a hypothesis is called an **experiment**.

If an experiment confirms a hypothesis, the experimental observations can add to the empirical basis of any original generalization. A generalization that has survived repeated experimental tests and is therefore supported by extensive observational evidence can be called a **theory** in the scientific sense. (Note that scientists are very careful about what they term *theories*! In everyday conversation we may designate all manner of ideas and opinions as *theories*, but these ideas could only be called *scientific theories* if they have a firm basis in observable fact.) On the other hand, if an experiment fails to confirm its hypothesis, the underlying theory may need to be reexamined and modified. Some theories can be based on both deductive and inductive arguments. For example, the theory of evolution, in addition to its basis in direct observations of biodiversity and genetics, also follows logically from general principles of population growth and resource limitation. This is one reason the theory is so robust.

The task facing any scientist is to relate generalizations and abstractions with concrete data. A successful scientist must be able to interpret data and search for broad principles – that is, to generalize – as well as plan experiments based on specific deductions or hypotheses.

Most of the exercises described in this book will direct you to make observations of well-described phenomena. This means that you will be employing empirical methods to verify facts for yourself. Furthermore, the activities are organized in a way intended to help you make important generalizations from your observations. These exercises are not labeled "experiments" however, as you will only occasionally be testing hypotheses. If you continue on in science, the observations that you make in this course will be a wealth of experience to draw upon as you are later called upon to formulate hypotheses of your own.

VARIABLES, CONTROLS, AND ALTERNATE EXPLANATIONS

The most important thing about any scientific investigation is that it be designed and carried out so that the results are clear and meaningful. As an example, let's suppose that you have isolated a new kind of bacterium and you want to see if it is killed by penicillin. For your test you place some of the bacteria in a suitable growth medium and add penicillin. Over the following week you observe no growth. Does this mean that the penicillin killed it? What if the temperature was wrong, or the medium incorrectly prepared – these factors could prevent growth, too. You want to be sure that the penicillin is the only **variable** being tested. (The word *variable* refers to something that varies and can be used in different scientific contexts. In an experiment, *variable* refers to any condition that the experimenter manipulates in order to explore its influence. Later in this chapter we will see how *variable* can also refer to a numerical quantity that is part of a graph or mathematical equation.)

In this example, the variable you are interested in is the presence of penicillin – does it influence the growth of your new bacterium? The easiest way to focus on this is to include a **control** in the experiment that is exactly the same as the test case except for the variable. You would grow two cultures side by side, one with penicillin and one without. If only the first culture did not grow, you could be pretty sure it was due to the penicillin, as it was the only difference between the two. (It's also usually a good idea to do several **replicates** of an experiment to make sure that your observations are verifiable.) In this case, the culture without penicillin is called a **negative control**, as it shows what would happen in the absence of the variable. You could also add a **positive control** to show how the variable acts when it is there. That is, you could also grow some organism that you already know is sensitive to penicillin. If it is not killed by the penicillin during the experiment, then you might wonder if your penicillin is any good. The proper use of controls in an experiment makes interpreting the experiment easier. It also helps you to figure out what went wrong if the results do not turn out as expected.

It is not unusual for people to get sloppy and omit controls from simple experiments that are easy to redo if problems occur. However, proper controls are critical for complicated, time-consuming, or expensive experiments. Some experiments are so complex that the negative controls are expected to change a little on their own. Such controls are often called **blanks**. The change in a blank must be subtracted from that found in the test case before the results can be read. Blanks are often used with spec-trophotometers and enzyme assays, as you will see in chapter 7.

GRAPHING AS A FORM OF GENERALIZATION

In order to maximize reproducibility, scientists usually make careful measurements of their observations. (Metric units of measurement used throughout this book are summarized on the inside back cover of the book; methods of measuring liquids and solutions are covered in appendix 2.) As a result, the information obtained from these observations, or **data**, are often in numeri-cal form. Two convenient ways to record and present numerical data are in tables and graphs.

A particular type of graph, the **line graph**, is more than just a way to display data. In a line graph the data points are connect-ed by lines in order to extrapolate what happens between the measured points. As such, the lines of a line graph are actually generalizations about what happens in cases that were not actu-ally measured, and can be used as the basis for deductions.

Figure 1.1 is an example of a line graph. It shows data from a hypothetical experiment in which proteins were separated by their relative size. As you will see in later chapters, proteins are huge molecules that can be anywhere from 100 to 100,000 times the size of a water molecule. It is possible to force mixtures of proteins through various types of gels in such a way that the smallest and most mobile proteins move the fastest, while the larger proteins lag behind. In this case, we have data from four proteins, with masses of 66,000; 44,000; 39,000; and 20,000 mass units. The smallest moves 3.6 cm through the gel, while the largest only moves 1.0 cm; the mid-sized proteins travel inter-mediate distances.

In this case the data are plotted to show how the distance trav-eled is a reflection of molecular size. By connecting the data points with straight lines, we can infer how other proteins might behave. For example, if we put a protein of unknown size on the gel and it moved 3.0 cm, we could use this graph to infer that the protein had a mass of about 26,000 units. In effect we have

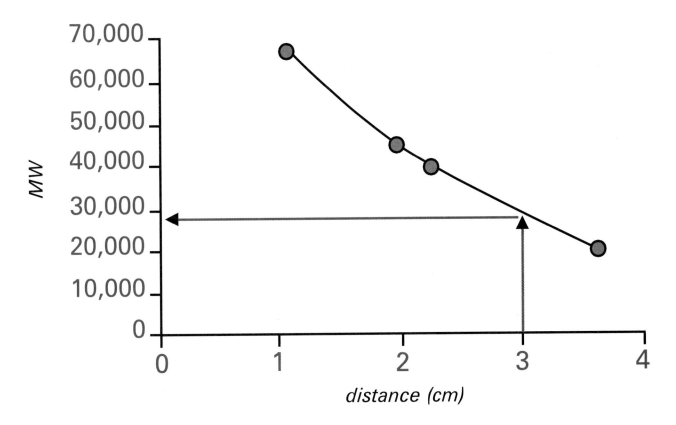

Figure 1.1 Separation of proteins by molecular size

standardized the gel for use as a measuring device for protein size. The process of standardizing an instrument or process for making measurements is called **calibration**, and a line graph used for calibration is called a **calibration curve**. This is an important scientific technique, and several exercises in this book will require you to construct calibration curves.

To be useful, a line graph must be constructed with an accurate scale for each axis – otherwise the interpolated lines will not be accurately calibrated. In figure 1.1, the horizontal axis is evenly spaced in centimeters to give a regular scale. As a result, the points plotted in the graph are not evenly spaced, reflecting the fact that the mid-sized proteins moved similar distances but were widely separated from the largest and smallest proteins. Likewise, the vertical axis is evenly spaced in units of 10,000 mass units rather than being skewed to emphasize the particular sizes of the proteins under study. Making graphs with improper scales is a very common error for beginners and you should take care to avoid this mistake.

In making any kind of graph it is important to use the two axes correctly. As you may recall from math class, the horizontal axis is called the x-axis and is used to plot the **independent variable**. (In figure 1.1 the independent variable is the distance moved by the proteins.) The vertical, or y-axis, is used for the **dependent variable**. How can you tell which variable is dependent and which independent? Once you set up an experiment, the independent variable is determined, and the value of the dependent variable will depend on what the independent variable is. In our example, the range of possible distances a protein might travel was fixed before the experiment began by the size of the gel and the system used to force the proteins through the gel. Once that is set up, the measurements of molecular weight we obtain depend on the distances we see the proteins travel. In general, if you can state your investigation in the format "What happens to Y as we change X?" then it follows that what you measure for Y will be plotted on the vertical axis of the graph and what you manipulate as X will be plotted on the horizontal axis.

EXERCISES WITH SCIENTIFIC LOGIC

The following exercises do not involve any actual measurements on your part – they are designed to focus on the analytical thinking skills you will use after measurements are complete. You will need to employ these skills in later exercises in order to understand the significance of your observations. (These situations are all fictional, but they are consistent with published data.)

A. Exploring generalizations and explanations

Observations of nature do not come with explanations – we must figure out what they mean by ourselves. The first explanation we may think of is not necessarily the best. The method of science is to examine various possible explanations for a phenomenon and then devise experiments to help us determine which explanation is the most accurate. For the following exercises, try to think of as many ideas as you can. There are no right or wrong answers, but some ideas will not survive careful scrutiny. There is nothing wrong with offering an idea that does not work out – a good scientist must always be ready to abandon an idea if the evidence does not support it.

1.   Review the Sherlock Holmes example from earlier in this chapter. Is Holmes' conclusion – that the crime was not committed by a stranger – the only possible explanation for the available evidence? What alternative explanations are there? How could a detective test these explanations to see which one is best?

2.   A public health doctor examines the situations of 800 men who died from lung cancer and finds that 96% of them smoked cigarettes.

- What is the variable being looked at in this study?
- What generalizations can you make from the data?
- How might you test these generalizations?
- Is it reasonable to guess from this study that similar results would be obtained if women were the subjects instead of men?

B. Planning an experiment with controls

Pretend that you are a food scientist working for a company that produces a flavoring syrup. The company is concerned about spoilage of the product and currently keeps it refrigerated to extend its shelf life. Your supervisor asks you to determine if any of three preservatives could be used instead of refrigeration. She gives you 10 days to perform the experiment but indicates that she is interested in any information you might obtain relevant to longer-term storage. Assume that someone has already developed a quick and easy test for determining if the syrup has gone bad or not, and that you can use this for monitoring your own samples.

1.   What is the variable you are testing? Make a list of what samples you would need to use to study this variable. What would be your control(s)?

2.   Within your experimental design, can you think of a convenient way to simulate long-term storage within your 10-day time frame? (That is, is there a way to make the syrup age at a faster rate than normal?) Expand your experimental design to include this treatment, with appropriate controls.

C. Reading graphs

A well-made graph can convey a great deal of information to someone who knows how to read it. Consider figure 1.2. This graph illustrates what happens when a solution of 3% hydrogen peroxide is mixed with a small amount of a substance called catalase, relative to a control that lacks catalase. In this experiment, the amount of hydrogen peroxide present in each solution was measured every three minutes after the addition of catalase.

1. What generalization(s) can you make about the effect of catalase on hydrogen peroxide?

2. Line a ruler up with the first three points of the line marked *catalase* and extend a straight line through those points until it intersects the x-axis.

- How does that line compare to the line marked *catalase*?
- Which line would represent a *constant* reaction rate?
- What does that tell you about the rate of catalase activity at time zero compared to time = 21 minutes? What can you conclude from this?

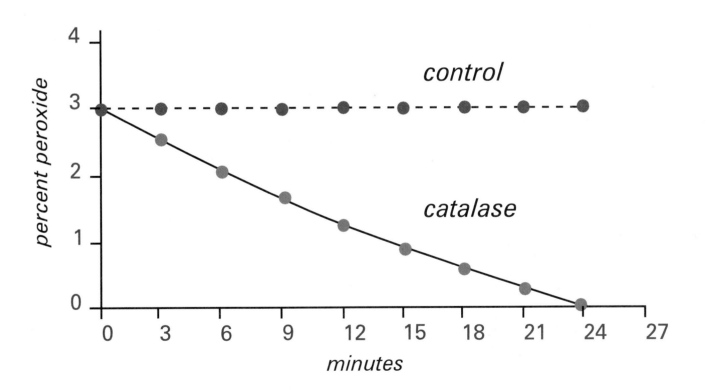

Figure 1.2 *Reduction in hydrogen peroxide concentration by catalase*

5

D. Constructing graphs from data tables

For each of the following data tables

- determine which is the independent variable and which is the dependent variable.
- construct a line graph that represents the data.
- state the generalization implied by the graph.

1. The following data represent what would happen if a protein solution were diluted to various concentrations (in mg per ml of solution) and each dilution measured to see how much ultraviolet light it absorbed:

Concentration	UV absorbance (in absorbance units)
15.0	2.80
10.0	1.92
8.0	1.51
5.0	0.94
3.0	0.56
0.0	0.00

2. The table below indicates the average daily intake of fat (in grams per day) for various countries along with the annual number of deaths due to breast cancer:

Country	Dietary fat	Cancer deaths per 100,000 people
Finland	115	13
Germany	130	17
Mexico	60	4
New Zealand	150	23
Spain	95	7
Thailand	25	1
United States	145	20

IMPORTANT TERMS

blank	deduction	independent variable	positive control
calibration	dependent variable	induction	replicate
calibration curve	empirical	line graph	reproducible
control	experiment	logic	theory
data	hypothesis	negative control	variable

Laboratory Report The Logical Basis of Science

FACTUAL CONTENT

Define the following terms. Brief definitions are given in the chapter, but you may also wish to consult your lecture notes and text for a better idea of how the terms are used. Pay close attention to those words that look similar but have different meanings – you will want to be able to tell them apart if they appear next to each other on a multiple-choice exam.

blank	independent variable
calibration	induction
calibration curve	line graph
control	logic
data	negative control
deduction	positive control
dependent variable	replicate
empirical	reproducible
experiment	theory
hypothesis	variable

EXERCISES *Record your data and answer the questions below.*

A. Exploring generalizations and explanations

1. Review the Sherlock Holmes example from earlier in this chapter. Is Holmes' conclusion – that the crime was not committed by a stranger – the only possible explanation for the available evidence? What alternative explanations are there? How could a detective test these explanations to see which one is best?

2. A public health doctor examines the situations of 800 men who died from lung cancer and finds that 96% of them smoked cigarettes.

    •   What is the variable being looked at in this study?

    •   What generalizations can you make from the data?

    •   How might you test these generalizations?

    •   Is it reasonable to guess from this study that similar results would be obtained if women were the subjects instead of men?

B. Planning an experiment with controls

Your company is concerned about spoilage of the product and currently keeps it refrigerated to extend it's shelf life. Your supervisor asks you to determine if any of three preservatives could be used instead of refrigeration.

1. What is the variable you are testing?

    Make a list of what samples you would need to use to study this variable. What would be your control(s)?

2. Within your experimental design, can you think of a convenient way to simulate long-term storage within your 10-day time frame? (That is, is there a way to make the syrup age at a faster rate than normal?) Expand your experimental design to include this treatment, with appropriate controls.

C. Reading graphs

Figure 1.2 illustrates what happens when a solution of 3% hydrogen peroxide is mixed with a small amount of a substance called catalase, relative to a control that lacks catalase.

1.   What generalization(s) can you make about the effect of catalase on hydrogen peroxide?

2.   Line a ruler up with the first three points of the line marked *catalase* and extend a straight line through those points until it intersects the x–axis.

   •   How does that line compare to the line marked *catalase*?

   •   Which line would represent a *constant* reaction rate?

   •   What does that tell you about the rate of catalase activity at time zero compared to time = 21 minutes? What can you conclude from this?

D. Constructing graphs from data tables

1.   In the space provided, graph the protein concentration against its UV absorbance.

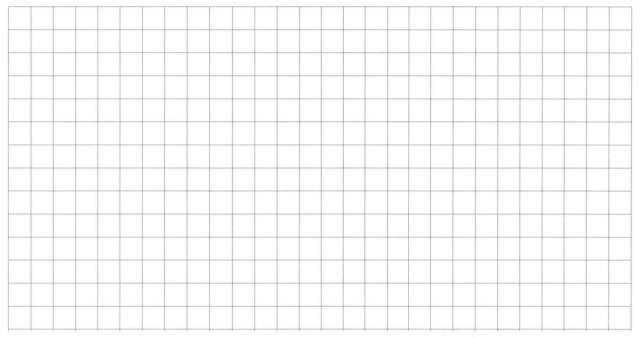

State the generalization implied by the graph

2.   In the space provided, graph the data for fat intake for various countries against the annual number of deaths due to breast cancer.

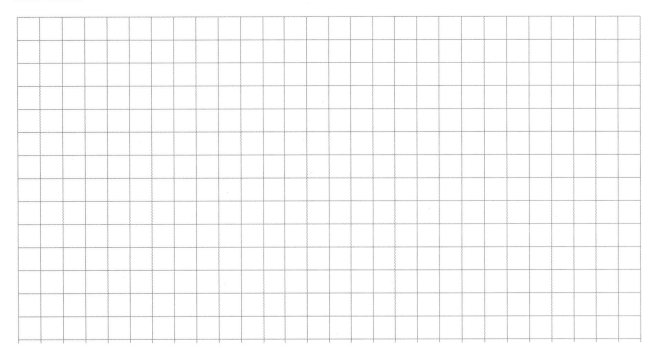

State the generalization implied by the graph

Water and Solutions

2

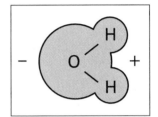

$$H_2O \longrightarrow H^+ + OH^-$$

$$Constant = [H^+] [OH^-]$$

$$H^+ + HCO_3^- \longrightarrow H_2CO_3$$

OBJECTIVES

New Skills

After working on the exercises in this chapter you will learn how to
1. make make serial dilutions of varying concentrations from a single stock solution.
2. measure the pH (acidity) of a solution using universal pH indicator.

Factual Content

In this chapter you will learn
1. to use terminology relating to solutions, acids and bases, and how materials move while dissolved in a solution.
2. some simple rules for using the standard pH scale.

New Concepts

In doing the exercises in this chapter you will observe how
1. molecular weight influences diffusion rates in a gel.
2. solutions of different concentrations may cause cells to swell or shrink.
3. buffers can protect a solution from changes in pH.
4. polar and nonpolar fluids interact and are affected by soaps.
5. nonpolar solvents affect cell membranes.

INTRODUCTION

Life as we know it is impossible without water. Living cells are at least 70 percent, and may be over 90 percent, water. Complex biological molecules are generally dissolved or suspended in water. In order to examine how water and solutions affect life, biologists have adopted many concepts and techniques used by chemists.

Some basic information on chemical measurement is collected in appendix 2 and on the inside back cover of this book. You should review this material thoroughly if you are not already comfortable with it from previous science courses. The rest of this chapter deals with general principles of chemistry that will be used throughout this book. The exercises will allow you to observe specific examples that illustrate these basic principles.

THE BASIC ELEMENTS OF BIOLOGICAL MOLECULES

The fundamental unit of an elementary substance is called an **atom**. Over 100 different elements are known, but only a few are important to life. These elements often combine to form new substances. A **molecule** is the basic unit of such a compound and is made of two or more atoms held together by chemical bonds. An atom contains a balanced number of negatively charged particles (electrons) and positively charged particles (protons) and therefore has a net charge of zero. However, it is not unusual for atoms and molecules to either lose or gain a few electrons. Those that gain electrons have a net negative charge, and those that lose electrons have a positive charge. Charged atoms or molecules are called **ions**. Ions of opposite charge tend to attract each other like tiny magnets; as long as they remain close together, they neutralize each others' electrical charges.

Roughly 80 percent of all elements are metals, but life is based on nonmetals. The nonmetallic elements most abundant in living material are:

hydrogen (H)   oxygen (O)   carbon (C)

nitrogen (N)   phosphorus (P)   sulfur (S)

None of these elements are ever encountered in their pure state in living systems; they are always bound together to form molecules and ions. Water is a compound of hydrogen and oxygen, so these two elements are the most common in biological systems. Carbon is the next most common element and is the basis for a bewildering variety of biological materials.

Several metals are also important to life. In living systems, metals always appear as positive ions. The most important metal ions are:

sodium (Na^+)   potassium (K^+)

calcium (Ca^{+2})   iron (Fe^{+2} or Fe^{+3})

Common negative ions include:

chloride (Cl^-)   hydroxide (OH^-)

bicarbonate (HCO_3^-)   nitrate (NO_3^-)

sulfate (SO_4^{-2})   phosphate (PO_4^{-3}).

Ionic compounds made of positive ions balanced by negative ions (for example, NaCl, $CaSO_4$, or K_3PO_4) are often called **salts**. Some salts, such as bicarbonate and phosphate, are important buffers.

DIFFUSION

Atoms, molecules, and ions are in constant motion. Because of this motion, the particles making up gases or liquids will tend to spread out and mix, a process known as **diffusion**. Diffusion in gases can be illustrated by the way a scent such as perfume will gradually drift outward from its source to fill a room; diffusion in liquids is exemplified by a drop of ink in a glass of water slowly dispersing its color throughout the water in the glass. Diffusion always occurs *from an area of high concentration to an area of low concentration*. In the example of the ink in the water, the ink molecules will generally move from the darkest, inkiest regions to the lighter areas (a few molecules might occasionally go from light to dark, but on the average, most go from dark to light). This is not because the individual molecules know where they are going, but because of probability: it is more likely that random movement will bring a molecule to a large open space than a small crowded space.

Diffusion is the process that allows substances to disperse in water and form solutions. Because of diffusion the various dissolved materials in a solution mix with each other and can engage in chemical reactions. This is as true inside a living cell as it is in glass of water. And since diffusion is governed by differences in concentration, biologists need to pay careful attention to the concentrations of the solutions they use (see appendix 2).

ACIDS, BASES, AND THE pH SCALE

An unusual feature of water molecules is that the electrons of the hydrogen atoms tend to congregate around the central oxygen atom, creating a slight negative charge on this side of the molecule. Naturally, this process makes the hydrogen atoms slightly positively charged.

Because a water molecule has a positive pole and a negative pole, it is said to be **polar** (figure 2.1). Many molecules are polar, but water is among the most polar. This polarity helps ionic compounds dissolve in water — as the oppositely charged ions diffuse away from each other, their electrical attraction to each other is lost somewhat amongst the electrical poles of the surrounding water molecules.

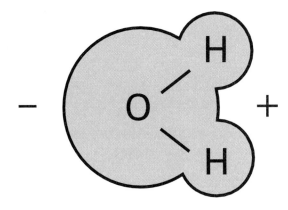

Figure 2.1 *Poles of a water molecule*

Because of the polar tension on the hydrogen-oxygen bond, it sometimes happens that the molecule "cracks" and a hydrogen pops off. Since the electrons were attracted to the oxygen, the hydrogen that comes off the molecule is ionized and positively charged.

$$H_2O \longrightarrow H^+ + OH^-$$

The tendency for water to dissociate into hydrogen and hydroxide ions is small. One liter of water contains 56 moles of water molecules. Of these 56 moles, careful measurements have shown that only 0.000 000 1 (or 10^{-7}) are dissociated at any given time. It is a natural property of water to maintain a constant equilibrium between its neutral and ionized forms such that only this tiny fraction of molecules dissociates.

However, substances added to water can change this equilibrium. An **acid** is defined as a compound that adds hydrogen ions to a solution. A **base** removes hydrogen ions from a solution. The simplest base is the hydroxide ion, because when it combines with hydrogen ions, it simply produces more water molecules. However, other ions and molecules can, and do, act as bases.

There are mathematical equations for describing solution equilibria. The derivation of these equations is best left to chemistry class, but we do need, at this point, to introduce the equation for the water dissociation equilibrium:

$$\text{constant} = [H^+]\,[OH^-]$$

What this means is that if you multiply together the concentrations of hydrogen ions and hydroxide ions, you will always get the same number, a constant. We already know that under nor-

mal conditions the concentration of each of these ions is 10^{-7} M. Therefore, the water constant is $10^{-7} \times 10^{-7} = 10^{-14}$. Now suppose that enough hydrochloric acid is added to make a 0.1 N solution. This means that the new hydrogen ion concentration is 10^{-1} M. The water molecules automatically compensate for this change by dropping the concentration of hydroxide ions to 10^{-13} M. That way, the product of the two concentrations remains 10^{-14}.

Because it is inconvenient to always write all of these negative exponents, chemists a long time ago devised the **pH scale** for describing acids and bases. pH is defined as the negative logarithm of the hydrogen ion concentration. Pure water has a hydrogen ion concentration of 10^{-7} M; the logarithm of this is –7 and the negative logarithm is simply 7. Therefore, the pH of pure water is 7. Similarly, the pH of the 0.1 N acid in the previous example would be 1.

Bases also affect the pH of a solution. Suppose you have a 0.1 N solution of NaOH. The concentration of hydroxide ions is 0.1 M or 10^{-1} M. From the water equilibrium equation we can calculate that this base drops the hydrogen ion concentration to 10^{-13} M for a pH of 13.

It is convenient to designate neutrality, pH 7, as the exact center of the pH scale. Thus the standard pH scale extends from 0 to 14. The more acidic a solution is, the lower its pH, and the more basic a solution is, the higher is its pH. Note that acids more concentrated than 1 N (pH 0) and bases more concentrated than 1 N (pH 14) are so strong that they go off the pH scale. Also remember that the pH scale is logarithmic – each pH value is 10 times more acidic than the next (i.e., pH 1 is 10 times as acidic as pH 2 and 100 times as acidic as pH 3).

To summarize:
   a.   The pH scale goes from 0 (acid) to 14 (base).
   b.   Neutral solutions are pH 7.
   c.   The pH scale progresses by powers of 10, with each greater value being 10 times less acidic and 10 times more basic than the previous value.

BUFFERS

When the hydroxide ion acts as a base and neutralizes a hydrogen ion, the product is water. When other bases accept hydrogen ions, the products are acids. For example, if the bicarbonate ion picks up a hydrogen ion, it turns into carbonic acid:

$$H^+ + HCO_3^- \longrightarrow H_2CO_3$$

Carbonic acid is an acid because it can lose a hydrogen ion by turning back into a bicarbonate ion. (Indeed, under the right circumstances, the bicarbonate ion is an acid, because it can lose a hydrogen ion and turn into the carbonate ion: CO_3^{-2}!) All this may seem confusing, but it is of crucial importance to biology. All of these molecules that switch back and forth between acids and bases act as **buffers**. A buffer helps stabilize the pH of a solution so that it does not change much if any extra acid or base is added to the solution. In exercise C you will observe the buffering effect for yourself. Even if you do not remember exactly how a buffer works, you will see that it is able to protect a solution from big changes in acidity.

Buffer molecules have ionization equilibria just like water does. At their equilibrium point, the buffer dissociates into equal concentrations of hydrogen ions and base ions, just as water does at pH 7. Each buffer has a particular pH where it reaches equilibrium. This is where its buffering capacity is strongest.

Most cells maintain a pH between 5 and 7. The proteins in the cell act as buffers to protect this pH. Many biological molecules cease to function properly outside of their native pH, so we must use many buffered solutions in the laboratory. Since buffers work best around their equilibrium points, you will see different buffers for different applications in these labs; for example, a phosphate buffer for work at pH 7 but an acetate buffer at pH 5.

ATOMS, COLOR, AND pH INDICATORS

The chemical reactions of a compound depend on the electrons of the atoms involved: electrons are shared to form covalent bonds, and are lost or gained to form ions. Physicists tell us that electrons are affected by electromagnetic forces. A stream of electromagnetic force is called electromagnetic radiation. There are many forms of electromagnetic radiation, but the most familiar is visible light: each color of the visible spectrum represents a different wavelength of radiation. If the electrons of an atom or molecule are arranged just right, they may be able to interact with different wavelengths of electromagnetic radiation. Those that interact with visible light appear colored. This principle is extremely important to biology, as it explains how chemistry and light can work together, as in photosynthesis or the reception of light by the eye. It also has practical implications in the laboratory, since the electron rearrangements in some chemical reactions are accompanied by color changes. Such reactions can be used for quick and easy chemical tests. In the exercises that follow you will use several colored compounds. In exercise C you will see how some dyes change color with changes in acidity as the positive hydrogen ions pull the negative electrons of the dye into new arrangements. Dyes that react to acidity are called **pH indicators**. Like buffers, each indicator works at its own particular pH. A mixture of several pH indicators that can be used over a wide range of pH levels is sometimes referred to as a universal pH indicator.

ORGANIC MOLECULES, POLARITY, CELL MEMBRANES, AND OSMOSIS

Molecules with a carbon backbone are called **organic**. Carbon atoms are the most versatile atoms on earth because they can each form four stable covalent bonds. These bonds are strong enough to build molecules thousands of atoms long without breaking. In combination with hydrogen, oxygen, and nitrogen, carbon is the basis of an infinite variety of compounds. We will see many of these compounds in later chapters.

Like water, some organic molecules are polar (some are even acids and bases). Other organic molecules are **nonpolar**. Nonpolar molecules do not interact well with polar molecules and actually repel water. As a result such molecules are often called **hydrophobic** (from the Greek *hydr-*, water, and *phobia*, fear or aversion, hence "aversion to water").

Lipids, a class of organic compounds that includes fats and oils, are all hydrophobic. The structure of fat and oil molecules is basically the same; fats are generally considered to be solid at room temperature while oils are liquid, but the distinction is not rigorous. All lipids conform to the general rule of chemistry that polar compounds only mix with other polar compounds while nonpolar compounds only mix with nonpolar compounds (that is, "oil and water do not mix" and "like dissolves like"). This property is very important to biology because cells use **membranes** made of lipid molecules to control diffusion.

If we think back to the diffusion example of an ink drop in a glass of water, it is obvious that the ink can diffuse throughout the water in the glass, but will not ooze through the glass itself. Cell membranes act as similar barriers. However, they are much thinner and more flexible barriers than glass, so they behave a little differently. First of all, oily substances can pass through the membrane simply by dissolving in the lipids that make up the membrane and diffusing across. Since oils do not dissolve well in the aqueous solutions inside and outside the cell, this does not happen very often, but it does happen. More importantly, tiny molecules like water, oxygen, and carbon dioxide pass through such membranes freely. They are smaller than the lipid molecules that make up the membrane and, as the membrane is not very thick, are able to slip between the lipid molecules almost as if the membrane were not there. Polar molecules larger than water are stopped by a lipid membrane, as are ions of any size. (This is because the charge of an ion attracts a "coating" of water molecules that make the ion seem larger than it really is — even a tiny hydrogen ion, when completely surrounded by water molecules, is too big to squeeze between the lipid molecules of the membrane.) Since some things can pass through a cell membrane, it is called a **permeable** membrane. Actually, it is **selectively per-meable**, as only certain things can pass through it.

Water is by far the most common substance in the environment of any cellular membrane and deserves special attention. Water molecules diffuse just like any other molecules do: from high concentration to low. In a solution, the concentration of water decreases as the concentration of solute increases, since the solute takes up space that would otherwise be full of water. A solution with a high concentration of solute therefore has a relatively low concentration of water. Living cells contain a solution of various compounds necessary for life. If those cells are placed in pure water, the water will diffuse through the membrane into the cell, because the cell has a low concentration of water relative to the solution outside of the cells. On the other hand, if the same cells are placed in a concentrated solution, water will diffuse out of the cell into the area of low water concentration. The diffusion of water across a membrane is called **osmosis**. You will examine the way solute concentration affects osmosis in exercise B.

When treated with strong bases such as lye (NaOH), fat and oil molecules are altered so that they pick up an ionic charge at one end. This means that one side of the molecule is nonpolar while the other side is very polar. As a result they can form a chemical bridge between oils and water, allowing them to mix somewhat. Fats and oils modified in this way are called **soaps**. (**Detergents** work the same way, but they are synthetic compounds and are not made from fats and oils.) As you will see, soaps and detergents can have dramatic effects on cell membranes.

EXERCISES WITH SOLUTIONS

REAGENTS

gel tubes	5% gelatin
dyes	0.01 M solutions of neutral red, methylene blue, phenol red, janus green, and/or aniline blue
NaCl stock solution	0.60 M sodium chloride
KCl solution	0.05 M potassium chloride
buffer solution	0.05 M sodium phosphate, adjusted to pH 7
protein solution	0.5% albumin
HCl	1 N hydrochloric acid
NaOH	1 N sodium hydroxide
universal pH indicator	commercial mixture of pH indicators
vegetable oil	
sudan III solution	95% ethanol saturated with sudan III stain
detergent solution	10% sodium dodecyl sulfate, or SDS
isotonic solutions	0.15 M solution of SDS and 0.3 M solutions of ethanol, isopropyl alcohol, sucrose, and glucose

A. Diffusion and molecular size

One of the factors that influences how fast a molecule diffuses is how massive it is—lighter molecules move more easily than heavy ones. The size of atoms and molecules is measured in mass units called Daltons. Water is a fairly small molecule with a mass of 18; common salts generally have a mass somewhere between 50 and 200, and organic molecules can range from a size smaller than water to thousands of times larger. In this exercise you will allow dyes of different molecular weights to diffuse through tubes of gelatin and measure what happens.

1.   Obtain one tube of gelatin for each dye available. Place one to two drops of a dye on top of each gel and label the tubes with the dye used. Set aside for 1 to 2 hours.

2.   After the dyes have had time to diffuse, measure how far the color has traveled in each gel. Measure from the top of the gel to the edge of the main region of dye. Ignore any thin streams of color that might indicate that the dye traveled through channels or other flaws in the gel. Record your data in a table in your report or notebook.

3.   Your instructor will give you the molecular weights of some, but not all, of the dyes. Graph the weight versus the distance traveled. Use your graph to estimate the weights of the unknown dyes. Compare your results to those obtained by other groups. How reproducible is this procedure?

B. Osmosis in potato cells

Osmosis refers strictly to the movement of water across a membrane. The direction the water moves depends on the concentration gradient across the membrane: which side has the lower water concentration and which the higher. It is difficult to change the contents of a cell without destroying it, but we can place it in media of different concentrations. A medium with a lot of solute and a water concentration lower than a cell is called **hypertonic**. It will make cells shrivel by osmosis because the water inside of the cells will be drawn out into the medium. Pure water and weak solutions are **hypotonic**. Hypotonic solutions cause cells to swell and sometimes even burst. Media that have the same concentration of water as cells are called **isotonic**.

1.   Obtain 200 ml of the NaCl stock solution. Make **serial dilutions** so that you have solutions of 0.60 M, 0.30 M, 0.15 M, 0.075 M, and 0.038 M NaCl. Do this by the following procedure:

a.   Place 100 ml of the NaCl stock solution in a beaker and set aside. This is 0.60 M.

b.   To the remaining 100 ml of NaCl stock solution, add 100 ml of water. Now you have 200 ml of 0.30 M NaCl.

c.   Place 100 ml of the 0.30 M solution in a beaker and set aside. Add 100 ml of water to the remainder to make 200 ml of 0.15 M NaCl.

d.   Continue this process to make 100 ml of 0.075 M NaCl and 200 ml of 0.038 M NaCl.

Also prepare a beaker of 0 M NaCl (pure water).

2.   Remove the skin from some potato slices and cut them lengthwise into strips until they are no more than 5 mm thick. Trim all pieces to the same length and record this length in your notes. Be sure to measure carefully to the nearest millimeter! The changes you will observe are small, so you will miss them if you do not measure precisely. Put three or four strips in each beaker of salt solution and wait 1 to 2 hours.

3.   Remove the potato strips from the beakers of salt solution (they are easier to handle if they are lightly rinsed and blotted dry with paper towels). Do not mix up the strips!

4.   Measure all of the strips and calculate the average length for each solution. Did the strips change length? Does the amount of change depend on the concentration of NaCl in solution? Tabulate your results in your notes.

5.   Construct a graph of your results showing the concentration of NaCl in each solution versus the corresponding change in strip length. From your graph can you figure out what concentration of NaCl is isotonic to potato cells?

C. Acids, bases, and buffers

Buffers are used to protect the pH of solutions from changing much if extra acid or base is added. You will now observe the buffering effect for yourself.

1.   Place about 5 ml of KCl solution in a test tube. KCl is *not* a buffer. Add a few drops of universal pH indicator to the solution until you have a definite color. Compare the color to the chart and determine the pH of the solution. Record the color and the pH value (and all subsequent observations) in your notes.

2.   Now add a drop of NaOH solution to the NaCl solution. Record the new pH.

3.   Add a drop of HCl and note the change in pH. Can you easily return the solution to its original pH?

4. Discard the solution and carefully rinse the test tube with distilled or deionized water. (This step is crucial to prevent contamination between solutions!)

5. Now place 5 ml of buffer solution in the test tube. This is a solution of a buffering base (i.e., phosphate) of the same molarity as the salt solution. It has been prepared so that it is at equilibrium. Add universal indicator and determine the pH.

6. Next add a drop of HCl and record the new pH. Add more HCl drop by drop until the pH changes two full pH units. Record the number of drops.

7. Discard the buffer, rinse the test tube, and add 5 ml of protein solution. Proteins are important molecules in cells and can act as buffers. Check the pH with universal indicator solution. Add the same number of drops of acid to this solution as you used in step 5. Is it as strong a buffer as the mineral buffer was?

D. Interactions between polar and nonpolar substances

Usually when lipids and water are mixed they quickly separate into two phases. All of the nonpolar molecules will stay together in one phase and all of the polar molecules will mix together in the other. Generally lipids can only mix with water if some kind of soap or detergent is present. A mixture of oil and water that does not quickly separate is called an **emulsion.** The common alcohol ethanol can act as a mild detergent because it is made of molecules with both polar and nonpolar ends.

1. Put 1 ml of vegetable oil and 4 ml of water in a test tube, cover the top, and shake to mix. Note the result in your report or notebook.

2. Add 5 to 10 drops of sudan III solution to the tube. Sudan III is a red dye dissolved in ethanol so it can interact with both the oil and the water. Shake the tube again and observe the result. Determine whether sudan III is polar or nonpolar.

3. Add 1 ml of detergent solution to the tube and shake again. Does an emulsion form? Let the tube stand undisturbed for a while to see how stable the result is.

E. Effect of nonpolar solvents on membrane lipids

Cells are held together by their lipid membranes, and the lipids in the membranes are held together because they will not mix with the water surrounding the cells. If the cells are placed in a solution in which lipids may freely dissolve, the membranes should be disrupted. In this exercise, blood cells will be placed in isotonic solutions of varying polarity. If the cells remain intact, the solution will stay cloudy (due to the cellular particles refracting light). If everything, including the cells, dissolves, the solution should become transparent.

1. Place 5 drops of sheep red blood cells in a tube containing isotonic SDS. Note the result. (Just look to see if the solution is less cloudy – any color change is due to other effects of SDS.)

2. Test the other solutions the same way. Explain your results based on the polarity of the solutions.

3. Look at your results from exercise B. How does the value for isotonic NaCl compare to the concentration of the solutions you used in this exercise? Explain.

IMPORTANT TERMS

acid	hydrophobic	molecule	polar
atom	hypertonic	nonpolar	salt
base	hypotonic	organic	selectively permeable
buffer	ion	osmosis	serial dilution
detergent	isotonic	permeable	soap
diffusion	lipid	pH indicator	
emulsion	membrane	pH scale	

Laboratory Report Water and Solutions

FACTUAL CONTENT

Define the following terms. Brief definitions are given in the chapter, but you may also wish to consult your lecture notes and text for a better idea of how the terms are used. Pay close attention to those words that look similar but have different meanings – you will want to be able to tell them apart if they appear next to each other on a multiple-choice exam.

acid

atom

base

buffer

detergent

diffusion

emulsion

hydrophobic

hypertonic

hypotonic

ion

isotonic

lipid

membrane

molarity

molecule

nonpolar

organic

osmosis

permeable

pH indicator

pH scale

polar

salt

selectively permeable

serial dilution

soap

What pH value is neutral?

Which end of the pH scale indicates acids?

Which end of the pH scale indicates bases?

How much more acidic is pH 4 than pH 5? Than pH 6?

EXERCISES - Record your data and answer the questions below.

A. Diffusion and molecular size

1. Tabulate the distance traveled by each dye and its molecular weight:

dye	molecular weight	distance

2. In the space provided, graph the molecular weight of each dye versus the distance traveled.

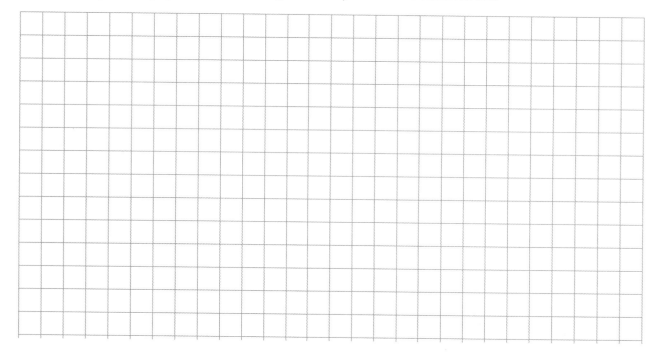

Use your graph to estimate the weights of the unknown dyes.

Compare your results to those obtained by other groups. How reproducible is this procedure?

B. Osmosis in potato cells

1. Tabulate the data from your experiment below:

NaCl concentration	length of potatoes before soaking	lengths after soaking	average length after soaking	change in average length

2. In the space provided, graph the change in length versus the NaCl concentration.

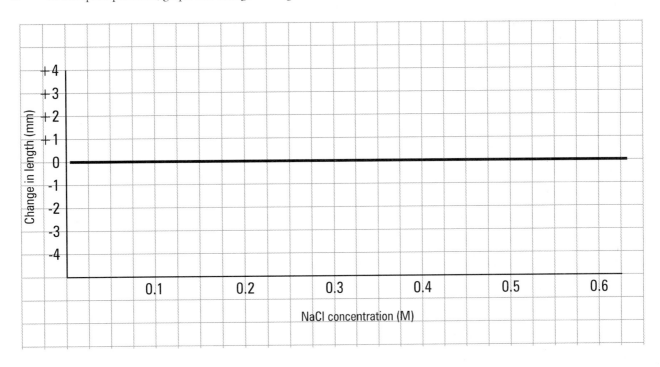

Does the amount of change depend on the concentration of NaCl in solution?

From your graph can you figure out what concentration of NaCl is isotonic to potato cells?

C. Acids, bases, and buffers

Record your data below.

1.   KCl solution – a non-buffer
     Record the color and the pH value of the solution mixed with indicator:

     Record the color and the pH value of the solution mixed with a drop of NaOH solution:

     Record the color and the pH value of the solution mixed with a drop of HCl solution:

     Can you easily return the solution to its original pH?

2.   Phosphate solution – a buffer
     Record the color and the pH value of the solution mixed with indicator:

     Record the color and the pH value of the solution mixed with a drop of HCl solution:

     Record the number of drops needed to change the pH two full pH units:

3.   Protein solution – a buffer?
        Record the color and the pH value of the solution mixed with indicator:

        Record the color and the pH value of the solution mixed with the same number of drops of HCl solution as you used to change the phosphate buffer two pH units:

        Is it as strong a buffer as the mineral buffer was?

D. Interactions between polar and non-polar substances

Record your observations below:

1.   Describe or diagram the appearance of the tube containing oil and water.

2.   Describe or diagram the appearance of the tube containing oil, water, and sudan III.

        Is sudan III polar or nonpolar?

3.   Describe or diagram the appearance of the tube containing oil, water, and detergent.

        Does an emulsion form? How stable is it?

E. Effect of non-polar solvents on membrane lipids

Tabulate your observations below:

solution	mixture cloudy or clear	solution polar or non-polar

Explain your results based on the polarity of the solutions.

Look at your results from exercise B. How does the value for isotonic NaCl compare to the concentration of the solutions you used in this exercise? Explain.

Cells and the Microscope

New Skills

After working on the exercises in this chapter you will learn how to
1. use a microscope to view cells.
2. estimate the sizes of objects viewed through a microscope.
3. prepare wet mount slides of specimens.
4. use colored stains to increase the contrast and visibility of particular structures of microscopic specimens.

Factual Content

In this chapter you will learn
1. the important parts of a compound microscope.
2. some basic terminology relating to optics.
3. to identify the parts of a eukaryotic cell that are visible with a light microscope.

New Concepts

In doing the exercises in this chapter you will
1. observe and compare some basic cell structures.
2. compare the size and structure of some prokaryotic and eukaryotic cells.
3. compare the size and structure of some animal and plant cells.

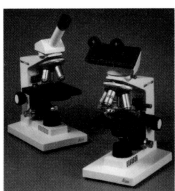

Photograph courtesy of Leica Microsystems Inc., Deerfield, IL

INTRODUCTION

Just as atoms are the basic units of matter, the basic units of life are **cells**. There are many types of cells, but as we shall see they can be sorted into two essential types. In today's lab you will observe several representative types of cells and learn something about their structure. These cells are very small, so you will have to use a **microscope** to see them.

THE COMPOUND MICROSCOPE

The existence of cells was totally unknown until people in Europe began experimenting with magnifying lenses a few centuries ago. The early systems of lenses were developed into more and more complex instruments. Today the most common type of microscope is actually a **compound microscope**, a system of lenses, mirrors, and prisms working together to produce high–resolution images (figure 3.1).

The most important parts of a compound microscope are the lenses. The viewer looks through the first lens, the **eyepiece** or **ocular**, at the top of the microscope. **Monocular** microscopes have a single eyepiece and **binocular** microscopes have a pair of eyepieces. At the other end of the body tube is a set of **objective** lenses. These are mounted in a revolving nosepiece so that one can select the objective offering the desired **magnification**. Objectives on a typical microscope include: scanning (typical magnification of 4**x**), low (10**x**), and high (45**x**). Some microscopes also have an oil immersion lens (100**x**) for very high magnification and resolution. The total magnification depends on the combination of both the ocular and the objective being used. The ocular usually has a magnification of 10**x**. This is multiplied by the magnification of the objective in place under the body tube, so the available range of magnification for the objectives listed above would be 40**x** to 450**x** (1000**x** for oil immersion).

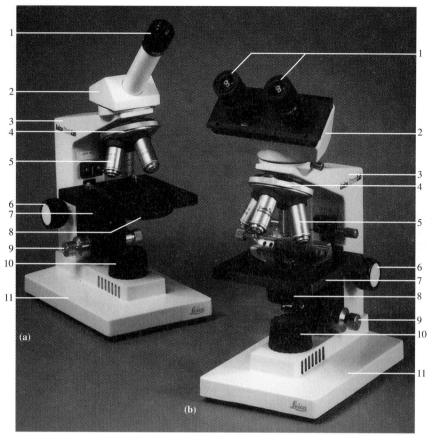

Figure 3.1 *(a) A compound monocular microscope and (b) a compound binocular microscope.*

1. *Eyepiece(s)*
2. *Body*
3. *Arm*
4. *Nosepiece*
5. *Objective*
6. *Focus adjustment knobs*
7. *Stage*
8. *Condenser with diaphragm*
9. *Condenser adjustment knob*
10. *Light*
11. *Base*

Beneath the objectives is an illuminated **stage** to support the glass slide holding the specimen. A **mechanical stage** includes a metal apparatus for gripping the slide and precisely positioning it under the objective. Focus knobs control the distance between the stage and the objectives, either by moving the stage or the body tube up and down. The **coarse focus knob** moves the stage or body tube the most, and the **fine focus knob** allows for more delicate control. The dimensions of the microscope are set so that the lenses are **parfocal**. This means that if you focus with one objective and then switch to another, the subject will remain in focus (or nearly so). You should always focus the specimen first using the scanning or low power objectives and then switch to higher magnification. Only the fine focus knob should be used with 45**x** or 100**x** objectives. These objectives are large, and using the coarse focus knob might make them crash into the slide!

Although many people concentrate on the magnification available with a microscope, a microscope's resolving power, or **resolution**, is also very important. Resolution is the ability to separate the images of adjacent objects. Resolution effectively measures how well the microscope can reveal fine detail. Resolution is not the same as magnification, because an image can be enlarged but still look so "fuzzy" that the details are lost.

One of the factors that influences resolving power is the refrac-

tion due to the material between the specimen and the lens. Light is bent (refracted) when it interacts with the electrons of the atoms of a material. Some materials refract light a great deal and so appear opaque to us. Transparent materials refract light to a lesser degree but still bend it to some extent. Air and glass bend light slightly differently, which is why a straw in a glass of water appears "bent" or "broken" when viewed from the side. When light from a specimen passes through a slide, it passes through air before it reaches the glass lens of the objective and therefore undergoes a similar kind of distorting refraction. All of this bending back and forth reduces the resolution of the system.

As it happens, mineral oil bends light in just the same way as glass. (A piece of glass in oil is invisible because of this.) Oil immersion lenses are designed to be used with a drop of oil placed between the lens and the slide, with the lens then focused close enough that it actually touches the oil. With this arrangement the light is not refracted but goes straight from the specimen to the lens. This increases resolution; without this increase in resolution, the high magnification of the lens would be worthless. If your microscope has an oil immersion lens you will find it very helpful in revealing the details of very small cells.

In order to see the details resolved by a microscope, we must usually enhance the contrast of the image. One way to do this is to control the width of the light beam coming into the specimen. For this purpose there is, beneath the stage, a device called the **condenser**. The condenser focuses the light onto the specimen. In some microscopes it can be moved up and down, but it should always be kept up all the way (that is, in contact with the stage) during use. The condenser includes a **diaphragm** to adjust the width of the light beam. If this is open too wide, the image will look bright but blurry. Each objective requires a different diaphragm setting, so every time you switch objectives you should adjust the diaphragm until the image is sharp and full of detail. Contrast is often further enhanced by the use of chemical **stains** that dye various parts of the specimen different colors and thus make them easier to differentiate.

OTHER TYPES OF MICROSCOPES

Dissecting microscopes are lens systems similar to compound microscopes but are simpler in several respects. They lack condensers and movable objectives, although many have knobs that adjust the internal optics to allow for different magnification levels. As a result, they cannot produce the same magnification and resolution as compound microscopes. On the other hand, dissecting microscopes are designed in such a way that they have a large working distance between the stage and body. They are used for examining the surface details of bulky specimens or for looking at specimens that must be manipulated while they are observed (such as dissections). You will not use a dissecting microscope in today's laboratory, but you will occasionally use them in future exercises.

Electron microscopes are yet another type of microscope. They are complex instruments that use beams of electrons instead of light to illuminate a specimen. Because of this they can magnify images tens of thousands of times with incredible resolution. They are too complex to use in a beginning course, but you will doubtless see many pictures taken with an electron microscope in your lecture textbook.

PROKARYOTES AND EUKARYOTES, THE TWO KINDS OF CELLS

Although cells are the fundamental units of life, they do contain smaller structures (just as atoms contain electrons, protons, and neutrons). There are many types of subcellular structures, but the easiest for beginners to see with a light microscope are those enclosed within their own membranes, the **organelles**. These include the **nucleus**, which contains the cell's DNA; **mitochondria**, involved in oxidizing sugar for energy; and **chloroplasts**, the sites of photosynthesis.

All cells have a cell membrane as their outer boundary, but not all cells have organelles and internal membranes. Those that lack any kind of organelle are described as **prokaryotic**. Cells that contain nuclei are called **eukaryotic**. Virtually all eukaryotic cells also have mitochondria, and eukaryotic cells that engage in photosynthesis have chloroplasts as well. Of the two types of cells the prokaryotes are far more common, as all bacteria are prokaryotic and bacteria live almost everywhere in large numbers. All other organisms are made of eukaryotic cells. Humans are included in this group, so eukaryotic cells hold a special interest for most people.

Not all subcellular structures are organelles. For example, most cells, including bacteria, plants, and fungi, but not animals, are encased in a rigid **cell wall**. The wall is outside the cell membrane and has no additional membrane of its own. Another important structure is the **chromosome**, a ribbon of DNA and associated molecules. Every cell has at least one chromosome. In eukaryotes these are located inside the nucleus; they are relatively large and can sometimes be rendered visible with appropriate stains. Prokaryotes generally have just a single chromosome whose ends are fused to form a continuous loop. Prokaryotic chromosomes are too small to be seen without an electron microscope.

In the following exercises you will examine some prokaryotic cells followed by a variety of eukaryotic cells. You may notice that the individual mitochondria and chloroplasts of the eukaryotic cells are very similar in size and shape to bacterial cells. These similarities parallel discoveries that indicate that mitochondria and chloroplasts have genetic and chemical affinities with bacteria. Indeed, it is now widely accepted that they are derived from bacterial cells — that is, the first mitochondria and chloroplasts were originally free-living bacteria that came to live inside of primitive eukaryotic cells whose only organelles prior to that were nuclei. This is one of the strangest events in the evolutionary history of life. *Symbiosis* is the term biologists use to describe

---

GENERAL GUIDELINES FOR USING THE MICROSCOPE

1. Carry the microscope with one hand beneath the base and one hand on the arm.

2. Keep all lenses clean. Use only lens paper to clean them.

3. Always focus by moving the lens away from the slide (upward).

4. Always focus with lowest power first. Center the specimen before going to a higher magnification.

5. After use:
   A. Switch to the lowest power objective;
   B. Remove the slide and lower the nosepiece;
   C. Return the microscope to its proper compartment.

an intimate association between two different kinds of organisms, and the idea that mitochondria and chloroplasts developed from symbiotic bacteria is called the **endosymbiotic theory**. We will examine some of the evidence for this theory in chapters 17 and 18.

EXERCISES WITH PROKARYOTIC AND EUKARYOTIC CELLS

You will be examining several preparations for the types of cells they contain. At times the process will be facilitated by using stains that selectively color certain cells or organelles. Be sure to draw pictures of what you see and answer the questions about the different cells in your notes—you will want this information for later reference.

A. Calibrating your microscope

1. Examine your microscope and look at the objectives carefully. The side of each lens should be stamped with its magnification. Make a table of these and calculate the total magnification for each objective by multiplying by the ocular magnification. (You can assume the ocular is 10**x** – the most common standard – unless your instructor tells you otherwise.) Note in your table if any of the lenses are special (such as oil immersion).

2. Set up your microscope with the scanning objective in place and adjust the light and diaphragm so that the field of vision is uniformly bright. Place a small plastic ruler under the objective and measure how wide the field of view is. Make another column in your table for "field width" and enter your measurement here.

3. Now switch to low power and remeasure the field. Since the field is more magnified, you should see less of it. For example, if the scanning objective is 4**x** and the low power is 10**x**, the low power field should be 4/10 the size of the scanning field. You cannot use a ruler on the high power objective, but its field can be calculated from the size of the low power field by comparing the magnifications. Using these field sizes, you may now estimate the size of everything you see under the microscope. (For example, if the low power field is 1.5 mm in diameter and it looks like it would take 10 cells to reach across the field, each cell must be about 0.15 mm wide.)

B. Prokaryotic cells

1. Place a drop of yogurt on a slide and cover with a coverslip. The "active cultures" in yogurt are bacteria whose activity sours milk and transforms it into yogurt. Search for some of these bacteria by examining your slide under the microscope. Begin with the scanning objective and focus on the white lumps of milk solids. Switch to low power and check your focus again. You will be unable to see bacteria at these magnifications, but these steps are necessary for establishing your focus and orienting yourself to the material on the slide. While viewing at low power, move the slide so that the edge of some solid material is centered in the field. This is the best place to find bacteria.

2. Now, *without moving the slide,* switch to high power. Adjust the diaphragm and focus as needed. Look for bacteria around the milk solids. Each bacterium is an extremely small rod or ball; often they stick together in chains.

3. Switch back to the scanning objective and remove the slide from the stage. Stain the yogurt by placing a drop of methylene blue next to the coverslip and drawing it under the coverslip with a piece of absorbent paper on the other side. Your lab instructor will demonstrate. Reexamine the slide by repeating steps 1 and 2. Methylene blue binds protoplasm (living material). Does this stain help you to find the bacteria? Does the stain reveal any structures inside the bacterial cell? Estimate the size of each bacterium and note this next to your sketch.

C. Nuclei and mitochondria in animal cells

1. Clean off your slide. Now you will look at some of your own cells. Gently scrape the inside of your cheek with a toothpick. Stir the scrapings into a small drop of water on your slide. Add a drop of methylene blue, cover with a coverslip, and examine under the microscope. Again, begin with the scanning objective. Move the slide until you can see a blue object in the center of the viewing field. Then switch to low power and adjust focus and diaphragm. Recenter the specimen and go to high power. Cheek cells look quite different from bacteria. Since most people have bacteria living in their mouths, you may also see bacteria near the cheek cells. How do your cells compare to bacteria with respect to size? Can you find the nuclei inside the cells? The area surrounding the nucleus is the **cytoplasm**.

2. Switch back to the scanning objective and remove the slide. Place a drop of janus green next to the cover slip and draw it under the coverslip with a piece of absorbent paper so that the cells are now stained with both methylene blue and janus green. Reexamine the slide quickly, as the janus green fades in about five minutes. Can you see dark spots sprinkled throughout the cytoplasm? These are mitochondria.

D. Chloroplasts and cell walls

1. Your instructor has prepared some slides of plant leaflets. Examine one of these as you did your cheek cells. Focus on the edge of the leaflet where it is only one cell thick. What do you see in the cytoplasm of these cells? These leaflets are unstained – the colored organelles are chloroplasts. Why were none visible in your cheek cells? Plant cells also have nuclei and mitochondria, but these are difficult to see without stains. You may have noticed

that the chloroplasts move around just inside the boundary of the cell, while the central area appears vacant. Many mature plant cells have a **water vacuole**, a membrane-bound reservoir of water, in their centers. The vacuole pushes the nucleus, mitochondria, and chloroplasts out toward the edge of the cell.

2.   Now make a slide of onion epidermis as demonstrated by your lab instructor. These are plant cells, but since they live underground they have no chloroplasts. Stain the onion with fast-green stain to highlight the thick cell wall. Did your cheek cells have a cell wall?

E. pH of cells

1.   Clean off your slide and make a new preparation of cheek cells. Instead of a stain, add a drop of universal pH indicator and cover with a coverslip. Examine under the microscope as you did in part C. (pH indicator is not a good stain and will not color the cells very intensely. Try your best to see what colors the cells turn – this may be easier under low power than high power.) What is the pH of the cytoplasm? Is the nucleus different? What does this tell you about the internal membranes of the cells?

2.   Now study a slide of onion epidermis made with universal pH indicator. How does the pH of onion cells compare to cheek cells? Do different parts of the cell show different pH values? How does that compare to the patterns you saw in the cheek cells?

F. Eukaryotic chromosomes

1.   Examine a prepared slide of onion root tip cells. These cells were fixed and stained as they were dividing to form new cells. Can you see a nucleus in the dividing cells? Can you find chromosomes? Were they visible inside the nuclei of the cheek cells? We will consider chromosomes in more detail in chapters 10, 12, and 18.

SUMMARY Comparison of Prokaryotic and Eukaryotic Cells

	Prokaryotic	Eukaryotic
cell wall	usually present	present in plants, fungi, and most algae
cell membrane	present	present
chromosomes	one tiny chromosome★	several (number depends on species), large
nucleus	absent	present
mitochondria	absent	present
chloroplasts	absent	present in photosynthetic cells
central vacuole	absent	present in mature plant cells
		Eukaryotic cells may contain other structures as well

★ Some bacteria have secondary chromosomes called plasmids.

IMPORTANT TERMS

binocular
cell
cell wall
chloroplast
chromosome
coarse focus knob
compound microscope
condenser

cytoplasm
diaphragm
dissecting microscope
electron microscope
endosymbiotic theory
eukaryotic
eyepiece
fine focus knob

magnification
mechanical stage
microscope
mitochondrion
monocular
nucleus
objectives
ocular

organelle
parfocal
prokaryotic
resolution
stage
stain
water vacuole

NAME

Laboratory Report Cells and the Microscope

FACTUAL CONTENT

Define the following terms. Brief definitions are given in the chapter, but you may also wish to consult your lecture notes and text for a better idea of how the terms are used. Pay close attention to those words that look similar but have different meanings – you will want to be able to tell them apart if they appear next to each other on a multiple-choice exam.

binocular

cell

cell wall

chloroplast

chromosome

coarse focus knob

compound microscope

condenser

cytoplasm

diaphragm

dissecting microscope

electron microscope

endosymbiotic theory

eukaryotic

eyepiece

fine focus knob

magnification

mechanical stage

microscope

mitochondrion

monocular

nucleus

objectives

ocular

organelle

parfocal

prokaryotic

resolution

stage

stain

water vacuole

A. Calibrating your microscope

Complete the table for your microscope:

objective	magnification	total magnification	field width
scanning			
low			
high			
oil			

B. Prokaryotic cells

Make a sketch of the bacteria as they appear under high power:

Did the methylene blue help you find the bacteria?

Does the stain reveal any structures inside of the bacterial cell?

Estimate the size of each bacterium.

C. Nuclei and mitochondria in animal cells

1. Make a sketch of the cheek cells stained with methylene blue as they appear under high power:

Label the nucleus and cytoplasm.

How do your cells compare to bacteria with respect to size?

2.   Make a sketch of the cheek cells stained with janus green as they appear under high power:

    Label the nucleus and mitochondria.

D. Chloroplasts and cell walls

1.   Make a sketch of the plant leaflets as they appear under high power:

    Label the chloroplasts and water vacuole.

    Why were no chloroplasts visible in your cheek cells?

2. Make a sketch of the onion cells stained with fast-green as they appear under high power:

Label the nucleus and cell wall.

Did your cheek cells have a cell wall?

E. pH of cells

1. Make a sketch of the cheek cells stained with universal pH indicator as they appear under low power:

Label the nucleus and cytoplasm. Next to each label note the color and pH.

Is the pH of the cytoplasm different than that of the nucleus?

What does this tell you about the internal membranes of the cells?

2.   Make a sketch of the onion cells stained with universal pH indicator as they appear under low power:

        Label the nucleus and cytoplasm. Next to each label note the color and pH.

        How does the pH of onion cells compare to cheek cells?

        Is the pH of the onion cytoplasm different than that of the nucleus?

        How does that compare to the patterns you saw in the cheek cells?

F. Eukaryotic chromosomes

        Make a sketch of the onion root tip cells as they appear under high power:

        Label the chromosomes.

        Can you see a nucleus in the dividing cells?

        Were chromosomes visible inside the nuclei of the cheek cells?

Carbohydrates and Polymers

4

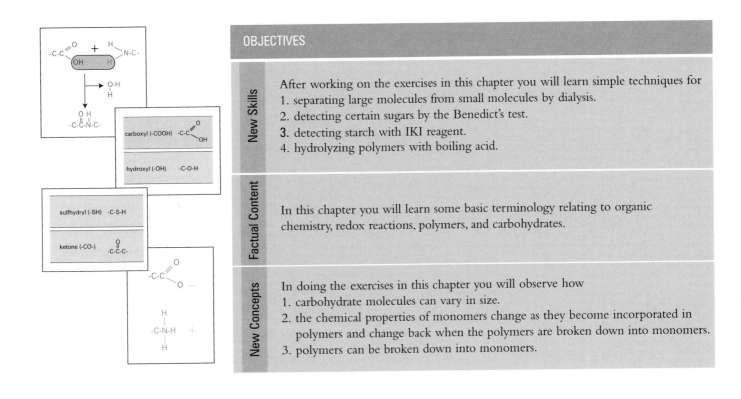

OBJECTIVES

New Skills

After working on the exercises in this chapter you will learn simple techniques for
1. separating large molecules from small molecules by dialysis.
2. detecting certain sugars by the Benedict's test.
3. detecting starch with IKI reagent.
4. hydrolyzing polymers with boiling acid.

Factual Content

In this chapter you will learn some basic terminology relating to organic chemistry, redox reactions, polymers, and carbohydrates.

New Concepts

In doing the exercises in this chapter you will observe how
1. carbohydrate molecules can vary in size.
2. the chemical properties of monomers change as they become incorporated in polymers and change back when the polymers are broken down into monomers.
3. polymers can be broken down into monomers.

INTRODUCTION

The organic molecules are the most interesting chemicals in a cell. They are also the most complex. Organic chemistry is a huge field of study in its own right, but in this chapter we can at least focus on a few principles of organic chemistry that are especially useful to biologists.

FUNCTIONAL GROUPS

As noted in chapter 2, there is an infinite variety of possible organic molecules. To make sense of them, chemists classify them by the functional groups they contain. A **functional group** is a particular arrangement of atoms with characteristic chemical properties. The groups most important to biology are diagramed in figure 4.1.

All except hydrocarbons are relatively polar groups and interact well with water. (Hydrocarbon groups are the most common functional groups in lipids.) In fact, carboxyl and amino groups actually ionize. Carboxyl groups are acidic (at least one is present in every organic acid) and lose hydrogen ions to form negative ions, while amines are basic and attract hydrogen ions to form complex positive ions (figure 4.2). This means that there is a complicated interplay between pH and the ionic charges organic molecules exhibit. As we shall see in the following chapters, this has many important implications.

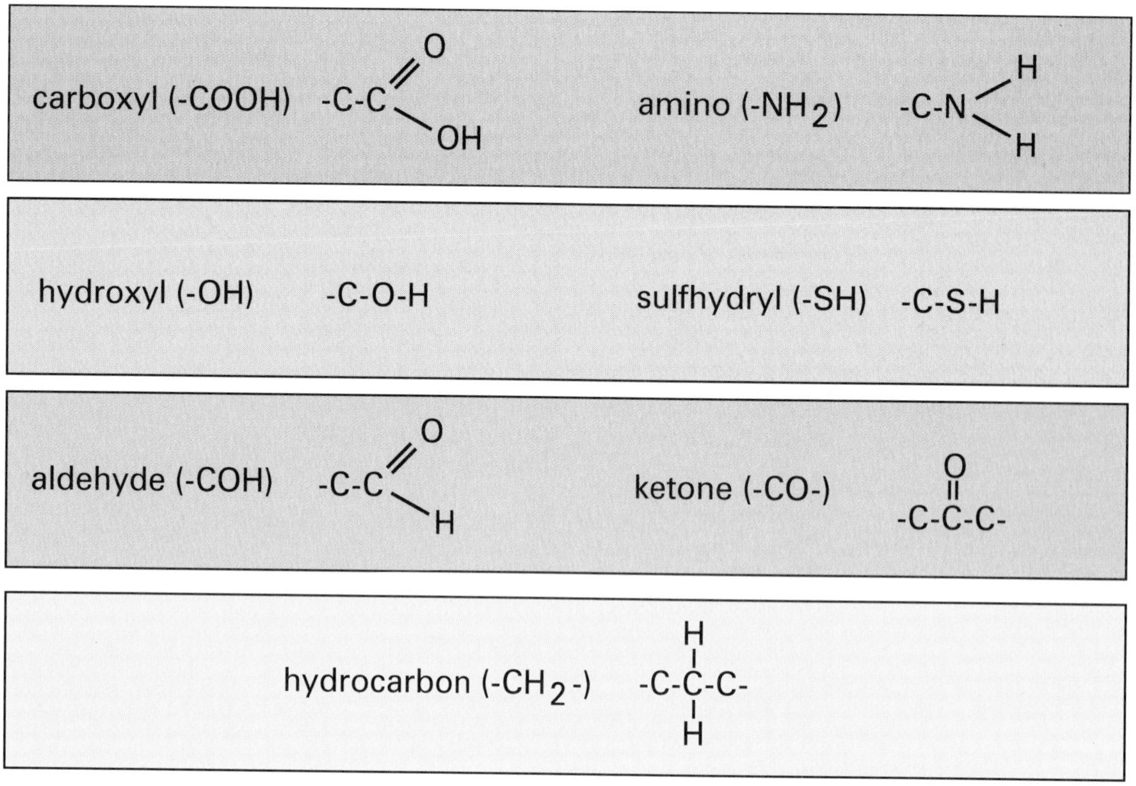

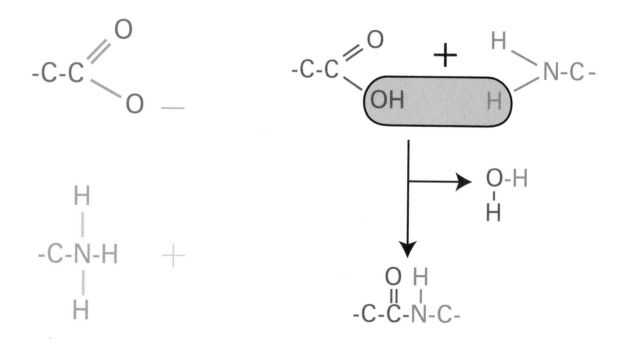

Figure 4.1 *Some common functional groups of organic molecules*

Figure 4.2 *Charges on organic acids and bases*

Figure 4.3 *Polymerization through water formation*

Large molecules contain many functional groups. A class of compounds has a particular combination of functional groups. We have already encountered one important class of organic molecules, the lipids. Other classes of compounds that are most important to biology are **carbohydrates**, **nucleic acids**, and **proteins**.

REDOX REACTIONS

In chapter 2 we said that chemistry depends on how the electrons are arranged around atoms. In some powerful chemical reactions, the electrons are not simply rearranged but actually jump from one atom or molecule to another. In a battery such reactions can create a steady stream of electrons, an electric current. When an atom or molecule receives new electrons from some other atom, it is **reduced** in charge. For example, a Cu^{+2} ion which gains two electrons becomes a neutral copper atom – its charge is reduced from +2 to 0. On the other hand, an atom or molecule that loses electrons will see its charge increase. Oxygen atoms are very good at stealing electrons from other atoms, so atoms that lose electrons are said to be **oxidized** (even if an atom other than oxygen is actually stealing the electrons – e.g., chlorine is a good oxidizer, too). Every time an atom is reduced, a nearby atom is oxidized – the electrons have to end up somewhere. Therefore, these reactions are called **oxidation-reduction** reactions, or **redox** reactions for short. As you might expect, these moving electrons mean that some redox reactions involve color changes, as when steel is oxidized into reddish-brown rust.

Redox reactions can release usable energy, as in a battery. The functional groups in organic molecules can be reduced or oxidized too, and these reactions can also release usable energy. In fact, the energy that powers your body is generated when the oxygen you breathe is used to oxidize the food that you eat. Like acid-base reactions, redox reactions are common in biochemistry, and we will see them again when we examine cellular metabolism. You may find both acid-base reactions and redox reactions easier to understand if you can imagine the electrons or hydrogen ions moving from one molecule to another.

POLYMERS

Within a particular class of compounds, some molecules have the minimum number of functional groups of their class while others are much larger. The smaller molecules with the fewer functional groups are usually quite sturdy and stable and are often able to link together in long chains to create giant molecules. Such molecular chains are referred to as **polymers**. The basic units that link up to make the polymeric chain are called **monomers**. It is much easier to break apart a polymer into monomers than to break apart a monomer into individual atoms. Large carbohydrates and all proteins and nucleic acids are polymers. Fats are built of smaller molecules too, but they are not truly polymers because the subunits are arranged in a parallel fashion rather than end-to-end in a chain.

Within a polymer, monomers are held together by chemical bonds. A common reaction occurs when an OH group on one monomer matches up with an H on another. When the two monomers are welded together, a water molecule is released (figure 4.3). This reaction is reversible – that is, if water molecules can be forced in between the links of a polymer, the polymer can be split up into monomers again. Such reactions are called **hydrolysis** reactions (from the Greek *hydr-* and *lysis,* meaning "splitting by water"). Hydrolysis can often be promoted by extremes of pH and/or high temperatures.

A polymer is more than a collection of monomers. When the monomers link together, they usually lose some of their chemical properties and gain new properties unique to the polymer. For example, the hydrocarbon ethylene is a gas at room temperature. When it is polymerized into polyethylene, it forms the soft plastic used for making plastic bags. Naturally, when a polymer is hydrolyzed into monomers, the polymeric properties disappear and the monomeric properties reassert themselves.

CARBOHYDRATES

A carbohydrate monomer is called a simple sugar or **monosaccharide**. All monosaccharides have one aldehyde or ketone group and two or more hydroxyl groups. Since these groups are polar, carbohydrates are generally water soluble. The only elements found in standard carbohydrates are carbon, hydrogen, and oxygen, present in the ratio 1:2:1. Thus the formula of any carbohydrate is a multiple of CH_2O (hence the name *carbohydrate)*. There are many different monosaccharides, but the most important have five or six carbon atoms.

Monosaccharides can polymerize to form larger molecules. A polymer of two simple sugars is called a **disaccharide**. Larger carbohydrate polymers are called **polysaccharides**. Polysaccharides can be several thousand monomers long and have enormous molecular weights. The most important polysaccharides are starch and cellulose (both commonly found in plants) and glycogen (common in animals and fungi). Some unusual polysaccharides found in the cell walls of algae, fungi, and bacteria contain additional functional groups not normally found in carbohydrates. Their structures will not be considered here.

In the following exercises you will use some simple chemical tests to detect sugars and starch. Both tests generate dramatic color changes. The test for sugars is based on a redox reaction; you can tell that the reaction involves a significant energy change because the reaction will need to be heated before the color develops. The test for starch is not a redox reaction and is instantaneous. You will use these tests in various contexts so that you can compare the properties of carbohydrate monomers to polymers.

EXERCISES WITH CARBOHYDRATES

Reagents

starch/ glucose solution	10% glucose/ 1% starch
monosaccharide solutions	1% solutions of glucose, fructose, and galactose
disaccharide solutions	1% solutions of maltose, lactose, and sucrose
polysaccharide solution	1% starch
Benedict's reagent	
IKI solution	
HCl	1 N hydrochloric acid

A. Separation of starch from glucose by size-exclusion dialysis

Dialysis tubing is an artificial membrane made from cellulose. Unlike a cell membrane, it is not made of lipids and exhibits no selectiveness based on polarity. However, there are gaps between the cellulose fibers that make up the membrane, and these gaps make the membrane permeable to some molecules. Because of this, dialysis tubing is often used as a model for cell membranes.

1. Obtain some dialysis tubing from your instructor. Moisten it and tie it into a bag as the instructor directs. Fill the bag with about 10 ml of starch/glucose solution. Tie the top of the bag so that the bag is loose and limp – not tight like a balloon. Rinse the outside of the filled bag with distilled water and place it in a large test tube of water. Wait at least 1.5 hours. (Do the remaining exercises while you wait.)

2. Remove the dialysis bag from the tube of water. Describe the appearance of the bag in your notes—is it still limp? Why?

3. Carefully open one end of the bag and divide the contents among two test tubes. Check one test tube with Benedict's reagent and the other with IKI using the procedures in exercises B and C. Record the results in your notes.

4. Now check the water outside of the bag by the same two tests. Explain your results.

B. Benedict's test for carbohydrates

The aldehyde or ketone group of a sugar is the most distinctive part of the molecule and can participate in several types of reactions. In basic solutions monosaccharides can be oxidized by Cu^{+2} ions. The copper ions pick up electrons in the reaction, changing from positively charged ions to uncharged metal atoms. Since copper ions are blue and copper metal is orange, this reaction is rather dramatic and makes an excellent test for detecting monosaccharides. This reaction is used so commonly for this purpose that it has a special name: the **Benedict's test**. You will now subject several six-carbon monosaccharides to the Benedict's test. Record your results in your report or notebook.

1. Mix 1 ml of 1% glucose solution with 2 ml of Benedict's reagent and place in a boiling water bath. Record the results.

2. Repeat for the other monosaccharide solutions (i.e., fructose and galactose).

Disaccharides may or may not give a positive Benedict's test — it depends on whether or not any aldehyde or ketone groups are still exposed after the monosaccharides are bonded together. Those sugars that do give a positive Benedict's test are called *reducing sugars*. All monosaccharides are reducing sugars.

1.   Test 1 ml of each disaccharide (i.e., 1% maltose, lactose, and sucrose) and polysaccharide (1% starch) with Benedict's reagent as you did with the monosaccharides.

2.   In your notes, construct a table of the different kinds of carbohydrates listing the results of the Benedict's test for each.

C. The IKI test

Some polysaccharides have a characteristic three–dimensional structure. This structure can form special complexes with iodine atoms to yield colored products. For example, starch has a coiled structure that turns blue in the presence of iodine, and glycogen, a highly branched molecule, turns red-violet.

1.   Put 1 ml of 1% starch solution in a test tube and add 1 to 2 drops of iodine/potassium iodide (IKI) solution.

2.   Test the mono- and disaccharides with IKI and record the results in your data table next to the results of the Benedict's test. Do you see a pattern in the results that depends on how polymerized the carbohydrates are?

D. Hydrolysis of a disaccharide

1.   Place 2 ml of sucrose solution in a test tube with 2 ml of HCl. Immediately take out 1 ml of the mixture and test with the Benedict's test. Record the result.

2.   Place the mixture in a boiling water bath and retest after boiling for one minute. How does the result compare to that obtained in step 1? Explain.

E. Hydrolysis of a polysaccharide

1.   Place 5 ml of starch solution in a test tube with 5 ml of HCl. Immediately take out one drop and test with IKI and take another 1 ml to test with the Benedict's test. Record both results.

2.   Place the mixture in a boiling water bath and redo both tests every 5 minutes. Can you observe the progress of the hydrolysis reaction?

IMPORTANT TERMS

Benedict's test	hydrolysis	oxidation-reduction	protein
carbohydrate	monomer	oxidize	redox
disaccharide	monosaccharide	polymer	reduce
functional group	nucleic acid	polysaccharide	

Laboratory Report Carbohydrates and Polymers

FACTUAL CONTENT

Define the following terms. Brief definitions are given in the chapter, but you may also wish to consult your lecture notes and text for a better idea of how the terms are used. Pay close attention to those words that look similar but have different meanings – you will want to be able to tell them apart if they appear next to each other on a multiple-choice exam.

Benedict's test

carbohydrate

disaccharide

functional group

hydrolysis

monomer

monosaccharide

nucleic acid

oxidation-reduction

oxidize

polymer

polysaccharide

protein

redox

reduce

EXERCISES - Record your data and answer the questions below.

A. Separation of starch from glucose by size-exclusion dialysis

Describe the appearance of the dialysis bag after you removed it from the tube of water – is it still limp? Why?

Complete the table for your results for the Benedict's test and the IKI test for the solutions inside the dialysis bag and outside the dialysis bag:

solution	Benedict's test	IKI test
inside bag		
outside bag		

Explain your results.

B and C. Benedict's test and IKI test for carbohydrates

Complete the table for your results for the Benedict's test and the IKI test for the carbohydrate solutions:

carbohydrates		Benedict's test	IKI test
monosaccharides	glucose fructose galactose		
disaccharides	maltose lactose sucrose		
polysaccharide	starch		

Do you see a pattern in the results that depends on how polymerized the carbohydrates are?

D. Hydrolysis of a disaccharide

Complete the table for your results for the Benedict's test on sucrose before and after hydrolysis:

solution	Benedict's test
before hydrolysis	
after hydrolysis	

How does the result after hydrolysis compare to that before hydrolysis? Explain.

E. Hydrolysis of a polysaccharide

Complete the table for your results for the Benedict's test and the IKI test during starch hydrolysis:

hydrolysis time	Benedict's test	IKI test
0 minutes		
5 minutes		
10 minutes		
15 minutes		
20 minutes		
25 minutes		

Can you observe the progress of the hydrolysis reaction?

Biological Macromolecules 1: Nucleic Acids and Proteins

5

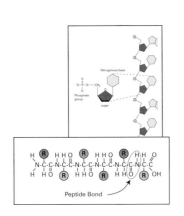

OBJECTIVES

New Skills	After working on the exercises in this chapter, you will learn how to 1. break apart eukaryotic cells and separate their organelles by centrifugation. 2. collect DNA by precipitation. 3. measure protein concentration by spectrophotometry.
Factual Content	In this chapter you will learn some basic terminology relating to the structure and chemistry of nucleic acids and proteins.
New Concepts	In doing the exercises in this chapter you will observe 1. where DNA is located inside a eukaryotic cell. 2. how protein solutions can be affected by heat, acid, and agitation.

INTRODUCTION

In the last chapter we introduced the concept of giant polymer molecules and used the polysaccharide starch as an example of a polymer. We saw that starch had some special properties due to its size (as shown by the IKI test) but that it could be broken down into much smaller monomers that exhibit the properties of monosaccharides (e.g., a positive Benedict's test). Very large polymers like starch are called **macromolecules**. In this chapter we will consider two important classes of macromolecules, **nucleic acids** and **proteins**.

As macromolecules go, starch is simple. That is because it is made of a single monomer, glucose, which is repeated in a chain hundreds of units long. That is, if you hydrolyzed a solution of starch completely, you would have a solution of the pure monosaccharide glucose. Nucleic acids are different. They are polymers of **nucleotides**, and every nucleic acid contains a mix of four different nucleotides. If you hydrolyze a solution of DNA, you do not get a pure solution of one kind of nucleotide, but a mixture of four different nucleotides. Proteins are even more complex – they are polymers of 20 different kinds of **amino acids**. Because of their complexity, nucleic acids and proteins are able to perform very special functions in living cells. They are central to understanding a cell's chemistry.

NUCLEIC ACIDS

There are two main types of nucleic acids: **DNA** and **RNA**. Both are polymers of nucleotides. A nucleotide molecule is rather complex, being made of a **nitrogenous base** molecule bonded to a "handle" composed of a sugar molecule and a phosphate ion. These "handles" link up in a polymeric chain in such a way that the nitrogenous bases stick out to one side, like the teeth of a comb. There are four different nitrogenous bases in any nucleic acid, and this accounts for the differences among the nucleotides – the phosphate parts are all identical, and the sugar is always the same in a particular nucleic acid: **ribose** in RNA and **deoxyribose** in DNA (figure 5.1).

Compared to other macromolecules, DNA and RNA do not take part in many chemical reactions. The most important thing about them is the sequence of the bases in each chain, because this carries the genetic code. The three-dimensional structure of DNA, from which we can now read the base sequence of individual genes, is quite complex. It was not worked out until the 1950s; its discovery earned Watson and Crick the Nobel prize. Only since about 1980 has it been possible to read the sequence of bases in a piece of DNA. This remains a difficult procedure suitable only for experienced scientists using complex equipment. In this lab we will focus on how DNA can be extracted from organisms so it can be studied. DNA does not hydrolyze well if it is simply boiled in acid, but we can use other techniques to fragment it once we have collected it. The techniques described here are a simplified version of what scientists must do each time they try to isolate a gene.

DNA is collected by the process of **precipitation**, which means forcing something to fall out of solution. Except when the cell is reproducing and the DNA condenses into distinct chromosomes, DNA is more or less dissolved in the cell's internal solution. If we change that solution, we can make it unable to hold the DNA anymore. Precipitation can happen to many macromolecules. If you have ever cooked a starchy food like oatmeal and then soaked the dirty pan in cold water, you may have noticed that the starch precipitates as a rubbery film in the cold water. Similarly, DNA can be precipitated with cold alcohol.

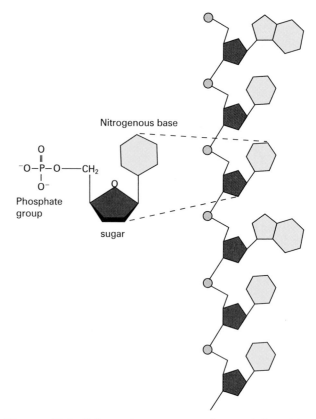

Figure 5.1 *Polymeric structure of a nucleic acid*

PROTEINS

Proteins are polymers of amino acids. An amino acid is an organic molecule about the same size as a monosaccharide, but its functional groups are quite different from those found in a carbohydrate. Each amino acid has an amino group at one end, a carboxyl group at the other end, and a third functional group of some kind in between. This third group can be one of several polar or nonpolar functional groups. They are known generically as **R groups**. Each of the twenty common amino acids has its own characteristic R group, which is what makes it different from all of the other amino acids.

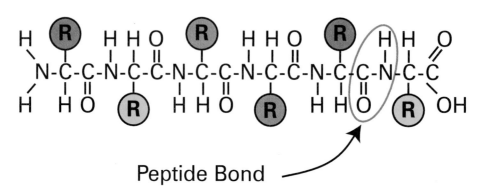

Peptide Bond

Figure 5.2 *Peptide bonds in a protein polymer*

Just as the sugar and phosphate parts of nucleotides join to form the polymeric backbones of nucleic acids, so the amino and carboxyl groups of amino acids link up to create protein polymers. The carboxyl group of one amino acid joins with the amino group of the next to create an amide, or **peptide**, bond. The backbone of any protein is a series of peptide bonds from which the R groups radiate out at right angles, as illustrated in figure 5.2.

Most proteins have several hundred amino acids, usually with at least one amino acid of each kind. With these hundreds of different functional groups, proteins are among the most complex molecules known. They can participate in a vast array of chemical reactions and control the chemistry of the rest of the cell. In fact, the technique you will use to hydrolyze the DNA you collect will involve exposing it to specific proteins that can cut DNA molecules.

The R groups within a single protein can interact with each other – nonpolar groups attracting other nonpolar groups, positively charged groups attracting negatively charged groups, and so on. As a result, the long backbone of a protein tends to fold up into a compact mass. The exact shape, or **conformation**, of the folded protein is determined by the amino acids it is made of and how they interact with the environment. Changing the environment (temperature, acidity, etc.) can change the conformation of the protein. A protein that becomes unfolded is said to be **denatured**. Denatured proteins often precipitate out of solution, which is what happens when milk curdles, for example. Some detergents can denature proteins while keeping them in solution. Proteins usually lose their biological activity when they are denatured, so they must be handled with care.

The ability of proteins to control chemical reactions makes them quite valuable. Proteins are routinely extracted and purified from all sorts of organisms, not just for research, but also for industrial purposes. In these situations it is important to be able to measure how much protein is present in the extract being used. While there are chemical tests for many of the different R groups, it is impractical to test a protein for each amino acid. There are, however, tests that detect the peptide bonds. In this lab we will show how one such test, used with an instrument called a **spectrophotometer**, can be used to measure the protein concentration of a solution.

EXERCISES WITH MACROMOLECULES

A. Nucleic acids: Extracting and digesting DNA

DNA is so complex that it cannot be easily synthesized in a laboratory. This means that it must be obtained from natural sources. This exercise will take you through the basic process of DNA extraction. At the end of the lab period you will save your samples for use in analytical procedures described in chapter 6.

Virtually all of the DNA in a eukaryotic cell is inside the nucleus. Therefore, the following procedure begins by homogenizing tissue in a buffer containing agents that are known to stabilize nuclei. The freed nuclei can then be separated from the other cell debris by **centrifugation**. A centrifuge spins a sample at high speed, resulting in centrifugal forces that are hundreds or even thousands of times the force of gravity. This makes small particles (in this case the cell nuclei) or undissolved materials settle to the bottom of the solution much faster than they normally would due simply to gravity. Material collected at the bottom of a centrifuge tube is called a **pellet**, and the liquid left on top of the pellet is called the **supernatant**. After the nuclei are recovered from the pellet, they will be treated with compounds that will destabilize the nuclei and release the DNA from the proteins that bind to it inside the nuclei. Since this is a multistep process, you should be prepared to note in your report or notebook any changes in your sample after each step.

Reagents

Extraction buffer	0.15 M sucrose; 5 mM $MgCl_2$; 0.1% NP-40; 50 mM NaCl; 10 mM THAM; pH 7.1
50 mM EDTA	
10% SDS	
2 M NaCl	
methylene blue solution	
ethanol	
DNA holding solutions	10% glycerol; 0.1% bromophenol blue; 0.1% orange-G; 10 mM THAM; pH 8.0; may have one of several possible nucleases added

1. Grind about 1 g of fresh wheat germ in a mortar with a pestle. Mix in 15 ml of cold extraction buffer. (The extraction buffer contains sucrose and NaCl so that it is isotonic to nuclei, as well as Mg^{+2} ions, which help stabilize the nucleus. It also contains a buffer and a mild detergent to dissolve the cell membrane.) Filter the extract through cheesecloth to remove debris, and place the filtrate in a centrifuge tube for centrifugation.

2. Give your centrifuge tube to your instructor to be loaded into the centrifuge. A centrifuge must be carefully balanced before it is used or the forces it generates will damage the machine. After centrifugation, discard the supernatant of buffer and cytoplasm and resuspend the pellet of nuclei in 5 ml cold extraction buffer. Keep this on ice until you are ready to work with it.

3. Before you lyse the nuclei, put a drop of the suspension on a microscope slide and stain with methylene blue stain. Can you find the individual nuclei?

4. Put 2 ml of the nuclear suspension in a test tube. Add 1 ml of 50 mM EDTA, mix gently, and let stand for 5 minutes. EDTA removes Ca^{+2} and Mg^{+2} ions from the solution and thus weakens the nuclear membranes.

5. Add 6 drops of 10% SDS solution, mix gently, and let stand 1 minute. What is the purpose of adding the detergent SDS?

6. Take a drop of your mixture now and make a new slide, staining as before. How has the sample changed?

7. Slowly and carefully add 6 drops of 2 M NaCl to your sample, mixing very gently after each drop. Note any changes.

The next procedure, while simple, is very delicate. If you are not careful, the ethanol will mix completely with the water. You must add the ethanol gently, so that it floats on top of the solution, and you must not agitate the preparation at all.

8. Take up about 2 ml of cold ethanol in a pipet (the exact amount is not critical). Slowly trickle the ethanol down the side of the tube. Once you see a definite layer forming on top of the DNA solution you may add the ethanol faster, but not so fast that the two layers begin to mix.

9. Take a long clean glass rod (It must stay clean! Hold it only by one end so your fingerprints don't contaminate the mixture!) and gently lower it into the tube until it penetrates both layers of solution. Carefully rotate the rod between your fingers and wind

the precipitating DNA onto the rod like spaghetti. Keep winding until you collect a visible mass of DNA. It should appear glassy, but it may be white if there is still some protein bound to the DNA. When you think you have all of the DNA you can recover, carefully remove the rod from the tube.

10. Put 1 ml of 2 M NaCl in a pyrex centrifuge tube and put the rod of DNA in the solution. Heat the tube in a boiling water bath to release the DNA into the solution.

11. Your DNA solution is reasonably pure, but it is dilute. To concentrate it, reprecipitate the DNA by adding 2 ml of cold ethanol to the tube and chilling on ice for 15 minutes. Centrifuge the tube to collect the DNA pellet.

12. Discard the supernatant. Add 50 µl of one of the holding solutions to your DNA. Some of these solutions contain proteins that will partially hydrolyze the DNA. Different people will use different solutions – follow your instructor's directions. Next week the different treatments will be compared.

13. Transfer the solution to a microfuge tube. Label this tube with your name and what kind of holding solution you used. Store as directed.

B. The biuret test for proteins: A quantitative analysis

In the **biuret test**, protein is exposed to copper ions in a basic solution. Although the test reagents are essentially the same as those used for the Benedict's test, the reaction is quite different. The biuret test involves not an energetic redox reaction, but the formation of a complex between the copper ions and the three-dimensional structure of the protein. Thus it is similar to the IKI test for starch. In this case, the copper ions line up with the regular pattern of equally spaced peptide bonds in the protein backbone to create a purple-colored complex. A useful aspect of the biuret reaction is that the darkness of the purple color is proportional to the amount of protein present.

The color of a solution can be measured with a spectrophotometer (figure 5.3). A spectrophotometer focuses a beam of light through a test tube onto an electronic photocell so that the amount of light passing through the test tube can be determined. The darker the solution in the tube, the more light will be absorbed. A prism-like device inside the instrument allows you to choose exactly which wavelength of light is focused on the tube, which means you can work with one color at a time. To provide reproducible measurements, a spectrophotometer must be used with a blank control.

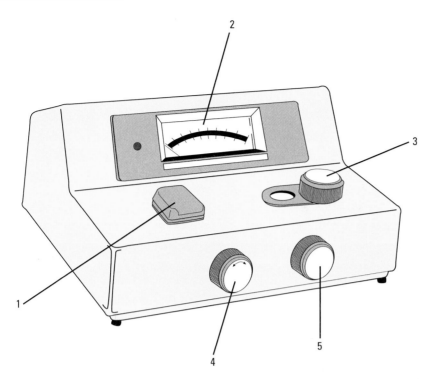

Figure 5.3 *A spectrophotometer*
1. *sample compartment*
2. *absorbance meter*
3. *wavelength selector*
4. *power knob*
5. *zeroing knob*

Reagents

albumin stock solution	2% (20 mg/ml) of bovine serum albumin or egg albumin in water
biuret reagent	0.15% $CuSO_4 \cdot 5\ H_2O$; 0.6% sodium-potassium tartarate; 3% NaOH

1. Obtain 3 to 5 ml of the albumin stock solution from your lab instructor. Make dilutions of this with distilled water so that you have samples of 20, 15, 10, 5, and 2 mg/ml.

2. Mix 1 ml of each sample with 4 ml of biuret reagent and let stand 30 minutes. (The color forms quickly but takes time to stabilize for measurement.) Also make a tube with 1 ml of water and 4 ml of biuret reagent to serve as a blank (0% protein).

3. Meanwhile, turn on the spectrophotometer and let it warm up. Make sure that it is set for a light wavelength of 550 nm. (While the spectrophotometer and solutions stand, you can do exercise C.)

4. Close the cover of the spectrophotometer while no tube is in the sample compartment. Adjust the power knob (on the left) until the meter reads "infinite absorbance" (infinite may be abbreviated by the symbol ∞). If you have trouble operating the instrument, ask your instructor for assistance.

5. Transfer solutions to the special spectrophotometer tubes for measuring.

HANDLE THESE TUBES CAREFULLY – THEY ARE EXPENSIVE.

First put the blank in the instrument and adjust the zeroing knob (on the right) to read "0 absorbance." Again, if you run into problems, ask your instructor for help.

6. Read the absorbance of each sample on the spectrophotometer. Tabulate your results.

7. In your notes, graph the absorbance of each sample versus the concentration of protein.

8. Obtain a sample of one of the "unknown" protein solutions from your instructor. Perform the biuret test on it and read the absorbance. Use your graph to figure out how much protein is in your unknown.

C. Denaturing proteins

Proteins are delicate and can easily be denatured by many things. Heat denatures most proteins, which is one of the reasons cooking changes food. For example, egg whites contain albumin, which turns into a white gel when heated. Egg whites also illustrate how simply agitating protein solutions can cause denaturation – that is what happens when they are beaten into a stiff foam. Other denaturants include acids, bases, detergents, strong salt solutions, and organic solvents. In the following exercise you will combine some methods to quickly denature egg albumin.

Reagents

albumin solution	2% egg albumin in water
acid	1 N HCl

1. Place 2 ml of albumin solution in a test tube and add 1 ml of acid. Heat in a boiling water bath until there is a visible change. What happened?

2. Place 2 ml of albumin in a clean test tube and seal with parafilm. Shake the tube vigorously 50 times (it should be foamy). Place in a boiling water bath for 5 minutes, then pour off the liquid. Can you find evidence of denatured protein? How does this compare to what you observed in step 1?

IMPORTANT TERMS

amino acid	deoxyribose	nucleotide	R group
biuret test	DNA	pellet	ribose
centrifugation	macromolecule	peptide	RNA
conformation	nitrogenous base	precipitation	spectrophotometer
denature	nucleic acid	protein	supernatant

Laboratory Report Biological Macromolecules I: Nucleic Acids and Proteins

FACTUAL CONTENT

Define the following terms. Brief definitions are given in the chapter, but you may also wish to consult your lecture notes and text for a better idea of how the terms are used. Pay close attention to those words that look similar but have different meanings — you will want to be able to tell them apart if they appear next to each other on a multiple-choice exam.

amino acid

biuret test

centrifugation

conformation

denature

deoxyribose

DNA

macromolecule

nitrogenous base

nucleic acid

nucleotide

pellet

peptide

precipitation

protein

R group

ribose

RNA

spectrophotometer

supernatant

EXERCISES - Record your data and answer the questions below.

A. Nucleic acids: Extracting and digesting DNA

Sketch the appearance of the nuclear suspension stained with methylene blue. Can you find the individual nuclei?

What is the purpose of adding the detergent SDS to the nuclear suspension?

How does the nuclear suspension, treated with EDTA and SDS, appear when it is stained with methylene blue and viewed under the microscope? How does this compare to the nuclear suspension before EDTA and SDS were added?

Record what kind of holding solution your DNA was stored in:

B. The biuret test for proteins: A quantitative analysis

Complete the table with the measured absorbance of each sample tested with the biuret test:

protein concentration	absorbance at 550 nm
0 mg/ml	
2 mg/ml	
5 mg/ml	
10 mg/ml	
15 mg/ml	
20 mg/ml	

In the space provided, graph the absorbance versus the protein concentration and draw a line through the points.

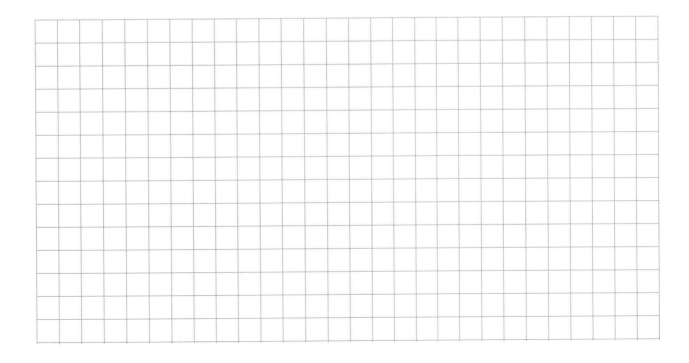

Record the absorbance of your unknown when tested with the biuret test:

Use your graph to estimate the concentration of protein in the unknown solution:

C. Denaturing proteins

1. Describe what happened to the albumin solution when it was treated with acid and heated:

2. Describe what happened to the albumin solution when it was shaken and heated. Can you find evidence of denatured protein? How does this compare to what you observed in step 1?

Biological Macromolecules II: Separation, Analysis, and Modeling

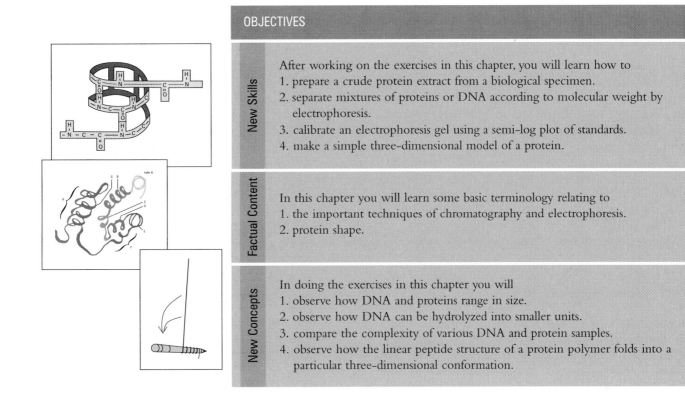

INTRODUCTION

In previous chapters you worked with some solutions that contained pure carbohydrates and proteins. The DNA extract that you prepared in the last chapter is not so pure. In all likelihood it contains proteins and other molecules that are bound to the DNA. Even if you were able to avoid these impurities, the extract would still contain a mixture of DNA molecules differing in size and nucleotide sequence. Preparing this kind of extract is usually just the beginning of a long process.

CHROMATOGRAPHY

One of the most important jobs of a chemist or biochemist is to separate complex mixtures into their various components for further analysis. For example, if you thought a plant extract might contain a drug useful for treating cancer, you would have to separate the therapeutic agent from all of the other materials in the original extract before you could produce a usable drug. The easiest separations to perform are those of colored materials, since you can watch them as they separate. An ink spot, for instance, if splashed with alcohol, will often spread out into rings of black, blue, and green dyes that together make up the pigments of the ink. Because many of the basic principles of chemical separations were originally worked out on colored mixtures, this branch of chemistry is called **chromatography** (from the Greek *chroma*, color). Traditional chromatographic separations are based on solubility – we separate molecules of differing degrees of polarity by adjusting the polarity of the solvent we use. Since molecules range from very polar to very nonpolar, with an infinite spectrum in between, the principle of solubility turns out to be quite flexible. In a few weeks we will use such a procedure to separate the various photosynthetic pigments in plant leaves.

In any chromatographic procedure, the sample is applied as a tiny spot near one end (the **origin**) of the separating matrix and forced through the matrix by solvent. As the process continues, the **solvent front**, the line between the wet matrix and the dry matrix, moves out farther and farther. When the solvent front is near the far edge of the matrix, the process is stopped and the position of the solvent front is noted. Each sample will leave a line of spots between the origin and the solvent front representing the individual parts of the mixture that have become separated. The distance traveled by a particular substance relative to the distance traveled by the solvent front is called the R_f; it should be constant for a given compound in a particular solvent-matrix system. That is, if a spot on a chromatogram is exactly halfway between the solvent front and the origin, it has an R_f of 0.5. If that spot is compared to a another spot run under the same conditions but showing an R_f of 0.7, it is almost certain that the two spots represent different compounds. In this way, chromatography can be used not only to separate the components of a mixture but also to identify unknown chemicals.

Chromatographic techniques can be used on macromolecules. Because of their size and complexity, macromolecules offer many properties on which to base separation techniques. One of the simplest methods is to sort them by size rather than solubility. This can be done in a tube of gel by a process similar to the method we used to study the diffusion of dyes in chapter 2. While such methods are used, they can take days. The technique of **electrophoresis** allows us to perform the same kind of separation in an hour or so.

ELECTROPHORESIS

In chapter 4 we saw that the presence of some functional groups – acids and amines – on organic molecules meant that these molecules could ionize and become electrically charged. Macromolecules have hundreds of functional groups and are capable of holding a considerable charge. In the case of a nucleic acid, the polymeric backbone is made of alternating units of sugar and phosphoric acid. Phosphoric acid is a strong acid, and at neutral pH it gives up most of its hydrogen ions to the surrounding solution. This means that the molecule has a charge of -1 for every monomer. The variety of R groups in proteins make them much more variable in charge; most proteins are somewhat negative at neutral pH, but that can change dramatically if the pH shifts.

Being electrically charged means that nucleic acids and proteins will move if subjected to an electric field. In electrophoresis, samples of nucleic acid or protein are placed in a gel between two electrodes. When power is applied to the electrodes, the negatively charged macromolecules will be repelled by the negative electrode and pushed through the gel towards the positive electrode. Obviously, any positively charged macromolecules will move the other way.

Because the charge density of a nucleic acid is always the same in a given buffer (i.e., -1 per monomer), size will be the only determinant in separating the various molecules in the sample. This means that in nucleic acid electrophoresis, R_f is proportional to molecular size. Protein electrophoresis is more complex, because the charge density of any particular protein is determined by the R groups of its amino acids, and the amino acid composition of each protein is unique. In many electrophoresis techniques the protein's R_f is determined by both charge and size, and it is impossible to tell much about either characteristic. Clinical labs employ such techniques when looking for specific blood proteins, and each protein is identified relative to a known standard rather than by its exact size or charge.

One method of protein electrophoresis that is very widely used is the technique called **SDS electrophoresis**. This technique is named after the detergent SDS, which is used to both dissolve and denature the proteins. In the process, the negatively charged SDS molecules surround and stick to the protein molecules, giving each a uniformly negative charge. Thus, SDS electrophoresis uses a kind of "chemical trick" to make the proteins all negatively charged so that they will separate by molecular weight the same way nucleic acid molecules do.

All kinds of electrophoresis use as a matrix some kind of gel made of an inert polymer suspended in a suitable buffer. For separating very large molecules, the gel is made of a chemically modified carbohydrate called agarose. This is useful for molecules with molecular weights from about 100,000 up to 10 million or so, a range that includes most nucleic acids. For molecules below

500,000, a denser gel is needed, and this is usually made of the synthetic compound polyacrylamide. (Sometimes "polyacrylamide gel electrophoresis" is abbreviated **PAGE**, and SDS electrophoresis is therefore referred to as **SDS–PAGE**.) Since the gel is always moist, there is no solvent front in the classical sense; instead a small dye molecule is added to the sample to create a **dye front** to mark how the separation is progressing.

THE THREE-DIMENSIONAL STRUCTURE OF PROTEINS

Proteins can contain anywhere from a few dozen amino acids (such small proteins are often called **peptides**) to 10,000 amino acids, with the average tending to be 400 to 500. Protein chains are unbranched – all of the monomers are in a line – so one would expect them to be like molecular ribbons. This is how they behave in an SDS gel, but physical measurements of unnatured proteins indicate that most are usually globular, as if the ribbon has become tied in a knot. This globular shape is the protein's conformation (cf. chapter 5). When a protein is in its normal shape, or **native conformation**, all of its R groups are arranged in their proper places. Specific clusters of R groups are what give the protein its biological activity. For example, every enzyme has a cluster of R groups called an active site (this will be discussed further in the next chapter). Denaturants take a protein out of its native conformation, which is why they destroy the protein's activity. Because of the link between protein conformation and function, biochemists are very interested in figuring out the exact shapes of the proteins they work with.

THE ALPHA HELIX

There are many reasons why a protein folds up in a particular way. One important factor is the way the R groups interact with each other, which is why the amino acid sequence of a protein is critical to its function (the amino acid sequence is what is carried in the genetic code). However, the repeating peptide bonds in a protein will cause a certain kind of folding no matter what the protein's sequence. That is because peptide bonds are polar – the oxygen atom in the acid group is slightly negative and the

hydrogen on the nitrogen atom in the amino group is slightly positive. They will attract each other in a weak **hydrogen bond**, and a series of hydrogen bonds will make the protein coil up into a helix (see figure 6.1). It takes about four amino acids to make one complete turn of the helix. The R groups are not directly involved in the helix, so they stick out from the coil like the threads of a screw. This structure was first worked out in proteins called alpha keratins that occur in hair and skin, so it was named the **alpha helix**. Linus Pauling worked out the exact dimensions of the alpha helix using carefully constructed scale paper models of peptide bonds. As this was the first step in understanding protein conformation, he won the Nobel prize in chemistry for his efforts. Since then, molecular model-building has become an important tool in biochemical research.

Most proteins are not pure alpha helix, since this would make them long tubes instead of compact knots. Globular proteins usually have several short coils of helix separated by lengths of uncoiled protein. These areas bend so that the coils can fold over each other into a compact mass. The amino acid R groups determine where the stretches of helix occur and how they fold in relation to each other.

MYOGLOBIN

One of the first proteins whose entire structure was completely worked out was myoglobin. Myoglobin is a small red protein found in vertebrate muscle cells. It is related to the hemoglobin that carries oxygen in blood and functions as a storage site for oxygen in the muscle.

Myoglobin happens to be a reasonably simple protein to model. That is because it is small for a protein (it has a molecular weight of about 15,000) and it is about 75% alpha helix. In exercise C you will construct a simple scale model of a myoglobin molecule. In the process you should see a dramatic change in size as a long, denatured protein attains its native conformation.

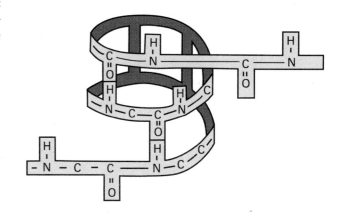

Figure 6.1 *The alpha helix*

Figure 6.2 *A horizontal gel electrophoresis system*

EXERCISES WITH ELECTROPHORESIS AND MODEL BUILDING

A. DNA electrophoresis

DNA is famous for being a double helix. This means that it is made of two polymeric chains twisted together like strands of yarn. As noted in chapter 5, a nucleic acid chain looks something like a comb, with the nitrogenous bases sticking out from the backbone like the comb's teeth. In the DNA double helix, the "teeth" of each chain face each other and meet in the middle, producing a shape like a ladder. As a result, each monomer is paired with a monomer on the opposite chain. Since each monomer has a nitrogenous base, the pairs are called **base pairs**, abbreviated as **bp**.

Most DNA molecules have molecular weights in the millions. Scientists have found it more convenient to express DNA size in terms of bp than Daltons. In this exercise you will calibrate your gel in bp using a commercially prepared standard of DNA fragments of known size. From this you will construct a calibration curve with which to measure your DNA samples from last week.

Careful studies have shown that for intensely negatively charged molecules, such as nucleic acids and SDS-complexed proteins, the observed electrophoretic R_f is inversely proportional to the *logarithm* of the molecule's molecular weight. By inversely proportional, we simply mean that the smaller the molecule, the farther it moves. That the relationship is logarithmic means, in practice, that the smaller molecules will be more spread out than the larger molecules. Constructing a calibration curve for a logarithmic relationship requires special semilogarithmic graph paper, which is included at the end of this chapter.

The DNA you collected last week is a collection of molecules of various lengths – we did not control for size. It is likely that there are so many different sizes that they may not separate cleanly into distinct spots. Instead, you will get a smear over a range of weights. The DNA will probably be fairly large and not move far from the origin, although the samples treated with nucleases should give smaller fragments. One of your tasks will be to compare what happens with the different treatments.

Electrophoresis in agarose is usually done in a horizontal gel submerged in a tank of buffer (figure 6.2). The origin is a series of wells indented in the side nearest the negative electrode. When the current is switched on, the DNA begins to migrate through the gel, but since it is not colored, it cannot be seen. For this reason the sample contains two dyes, bromophenol blue and orange-G. Under these conditions, these dyes migrate at the same rate as DNA fragments of about 250 and 70 bp, respectively. The dye molecules are not really that large, but since they are not as heavily charged as DNA, they move relatively slowly. The orange dye will be used as the tracking dye for calculating R_fs. Once the electrophoresis is complete, the gel will be treated with a chemical to make the DNA visible. The most commonly used compound for this is ethidium bromide, which, when in contact with DNA, glows pink under ultraviolet light. Because ethidium bromide attaches to DNA, it is very poisonous and is believed to cause genetic mutations and cancer. Safer, colored stains can also be used to visualize DNA, but these methods are not as sensitive as ethidium bromide. Your instructor will tell you what method the class will be using.

Reagents

electrophoresis buffer	0.05 M THAM; 2 mM EDTA; acetic acid to pH 8.0
agarose	1% agarose in electrophoresis buffer

1.   Your instructor may have already prepared an agarose gel for you to use. If not, each group should prepare a clean plastic casting tray by sealing both ends tightly with tape.

2.   Place a tube of agarose in a boiling water bath until the agarose melts completely.

3.   Pour the agarose into the gel tray. Gently rock the tray back and forth once to distribute the agarose as evenly as possible. Put the well-forming comb in the slots in the side of the casting tray and push the comb down into the gel until it stops. Allow the gel to solidify.

4. When the gel is completely firm, carefully remove the well comb and the tape from the ends of the tray. Place the gel in the electrophoresis chamber as your instructor directs. The wells must be on the side with the negative (black) electrode.

5. Fill the chamber with electrophoresis buffer until it covers the gels by 2 to 3 mm.

6. Observe as your instructor pipets 10 to 20 µl of a calibration standard into one of the wells. The sample contains 10% glycerol, so it is denser than the electrode buffer and should sink to the bottom of the well. This is easy to see because of the dye in the sample. Use the same technique to load another well with 20 µl of your DNA sample. Because a gel has several wells in it, other students will load their samples in the wells next to yours. Be sure to record in your notes what each well contains.

7. Put the top on the chamber. For safety, the chamber is designed so that it will not work without the top. Connect the chamber to the high-voltage power supply and run as your instructor directs.

8. Run the gel until the orange dye is within 1 cm of the positive end of the gel. It should have separated from the blue dye.

9. At the end of the electrophoresis, turn the power off and disconnect the chamber from the power supply.

10. Remove the gels. Measure the distance between the dye spots and the sample well (the origin).

11. Place the tray in cold water for a minute so that the gel becomes firm. If the gel does not float off the tray, carefully loosen it with a metal spatula while it is still submerged. Transfer the gel to a plastic storage box and stain as your instructor directs. Label the side of the box with your name.

12. After staining is complete, the gel should exhibit colored bands where DNA is present. Are the patterns as expected? Did the amount of nuclease you added to the DNA affect the size of the fragments?

13. Measure the distance from each band to the origin and compare these to the distance traveled by the dye front. Record the distance and calculate the R_f by dividing the distance traveled by the DNA by the distance traveled by the dye front. Record the R_f for each band. (The R_f should always be a fraction between 0 and 1.)

14. Copy the semilog paper at the end of this chapter to make a calibration curve. Notice that the vertical scale is not linear – it periodically "bunches up" as you move up the scale. That is because the distance between the lines corresponds to the *logarithm* of the difference between the values of the vertical scale. That is, unlike a linear scale, where the distance between 1 and 100 is ten times the distance between 1 and 10, here the distance between 1 and 100 is twice the distance between 10 and 100 (because the log of 1 = 0, log 10 = 1, and log 100 = 2). Use this vertical scale to plot numbers of base pairs from 100 to 100,000.

15. Plot R_f across the bottom on the linear scale, from 0 to 1.0

16. Graph the base-pair numbers given you by your instructor for the standards against the R_fs you observed on your gel. Draw the best straight line you can through the points. This is your calibration curve.

17. Use your calibration curve to estimate the number of base pairs in each DNA band on your gel (or, if no distinct bands are present, the high and low ends of the smeared range of DNA sizes).

B. SDS electrophoresis of proteins

As SDS electrophoresis samples are dissolved in a detergent solution, the sample buffer works very well at extracting proteins from tissue. For this exercise you will be given a choice of different biological specimens from which to make samples.

Acrylamide gels are run in a vertical apparatus with the gel sealed between an upper and a lower buffer tank (figure 6.3). Otherwise the procedure is very similar to DNA electrophoresis: A standard mixture of proteins is used to calibrate the gel, the proteins are made visible with a colored stain, R_fs are calculated relative to a dye front, and the data are plotted on semilog paper.

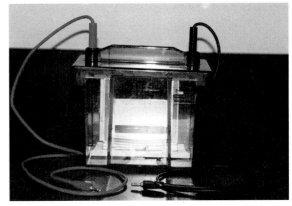

Figure 6.3 *A vertical gel electrophoresis system*

SDS-PAGE can in some circumstances tell you more about a protein than its molecular weight. That is because many proteins are actually complexes of subunits. For example, the red blood protein hemoglobin is actually made of several smaller proteins. In many cases the subunits can be detached from each other by the reagent mercaptoethanol. Thus, if a protein sample is prepared two ways, with and without mercaptoethanol, a comparison of the resulting gel patterns can indicate what the subunit structure of the protein is like. In the case of hemoglobin, a sample run without mercaptoethanol gives a weight of 60,000, whereas a sample reduced with mercaptoethanol gives a single band of 15,000. Since it takes four subunits of weight 15,000 to make a protein of weight 60,000, these data indicate that there are four subunits of identical weight in hemoglobin.

Reagents

electrophoresis buffer	0.2 M glycine; 0.025 M THAM; 0.1% SDS; pH 8.3
sample buffer	0.0625 M THAM; 2% SDS; 10% glycerol; 0.025% bromophenol blue; pH 6.8; with or without 0.1% mercaptoethanol
protein samples	standards and unknowns in sample buffer, approximately 0.1 to 1 mg/ml
stain	0.025% coomassie blue G-250 in 10% acetic acid, or equivalent

1. Acrylamide is a neurotoxin that must be handled carefully. Your instructor will prepare the gels and assemble them in the vertical apparatus for you.

2. Choose one of the tissue types available and collect about 1 to 2 mg of animal tissue or about 100 mg of plant material. (This is equivalent to 1 to 2 fruit flies or a square of leaf 2 cm on a side.)

3. Mash the tissue in a microfuge tube with the end of a toothpick. Add 100 µl of sample buffer and place the tube in a boiling water bath for 5 minutes.

4. Remove the tube from the water bath and centrifuge to pellet out the debris.

5. Observe as your instructor loads 10 µl of protein standard in one of the wells of the gel. Use the same technique to load 10 µl of your sample to the gel.

6. Your instructor has several different proteins dissolved in SDS buffer. These are purer than the samples you have made, and your task will be to determine the size of the proteins in each sample. Load 10 µl of these in other wells. Be sure to record in your notes what samples you use and which well each one is in.

7. Put the top on the chamber. For safety, the chamber is designed so that it will not work without the top. Connect the chamber to the high-voltage power supply and run as your instructor directs.

8. Run the gel until the dye front is within 1 cm of the positive end of the gel.

9. At the end of the electrophoresis, turn the power off and disconnect the chamber from the power supply.

10. Remove the gels, being sure that you only handle the gels with gloved hands. Measure the distance between the dye front and the origin.

11. Place the gel in a plastic storage box and rinse with distilled or deionized water. Cover the gel with stain and label the side of the box with your name.

12. Two to three hours later, pour off the stain solution and cover the gel with water. Change the water after 24 hours. The water will need to be changed several times in order to wash out the background stain — your instructor may take care of this for you.

13. After destaining, the gel should exhibit colored bands where protein is present. Measure the distance from each band to the origin and compare to the distance traveled by the dye front. If the dye front is no longer visible, use the measurement you recorded in step 10. Record the distance and the R_f for each band.

14. Use the semilog paper at the end of this chapter to make your calibration curve. Use this vertical scale to plot molecular weights from 10,000 to 1,000,000.

15. Plot R_f across the bottom on the linear scale, from 0 to 1.0

16. Graph the molecular weights given you by your instructor for the standards against the R_fs you observed on your gel. Draw the best straight line you can through the points. This is your calibration curve.

17. Compare the R_f of your unknown samples to your calibration curve to determine the molecular weight of your unknown proteins. Compare the size of your proteins to a water molecule (MW = 18) and a glucose molecule (MW = 180).

18. Examine the bands from your tissue samples. How many bands are there? (Many tissues will give numerous bands.) How do the different tissues compare? Are there any particularly dominant bands? What size are they?

C. Modeling myoglobin structure

Myoglobin has a total of eight coils of alpha helix of varying sizes. They are designated A through H, starting at the end of the protein with a free amino group (the N-terminal end). The protein has hairpin turns of five to eight amino acids between some helical segments.

1. Obtain a length of wire 1 meter long. This will represent the peptide backbone of the myoglobin polymer. R groups would radiate out from this like the barbs on barbed wire, but we will not concern ourselves with them in this model. On this scale, every 2 cm of wire represents about three amino acids. The measurements given in these instructions correspond to this scale – you may have to adjust them slightly, depending on how tightly you wind your initial coil.

2. Begin wrapping the wire around a pencil, leaving a 1 cm "tail" at the end. Wind the wire clockwise as you extend the coil away from you (figure 6.4). Continue winding until the entire wire is "alpha helix." Measure the length of the coil and record the length in your notes. You should have reduced the length of the wire about 75%.

3. Helix A is only four loops long, so count four loops of the coil. Helix B follows immediately, so bend the coil at this point into a 90° angle to the left and count out another four loops for B. Again helix C follows immediately after B, so bend the coil 120° toward you for two more coils (figure 6.5).

4. After the second coil of C, pull the wire to straighten out a 5 cm length between helix C and helix D. Bend this into a hairpin loop to the right and begin the two coils of helix D. It should be above helix C and oriented 90° forward from it (figures 6.5 and 6.6).

5. Helix E follows immediately. Bend the coil 90° straight down for four more coils. Helix E should run parallel to helix A.

6. At the end of helix E, straighten out another 5 cm segment and make a hairpin loop to the left. Follow this with two coils for helix F.

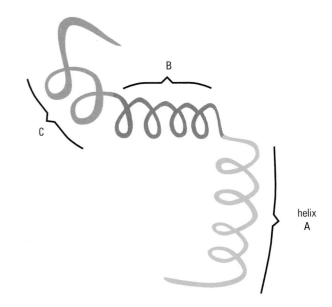

Figure 6.5 *Arrangement of coils A to C (side view)*

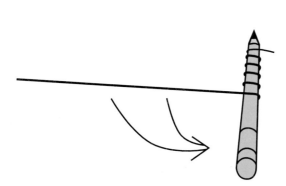

Figure 6.4 *Winding a coil*

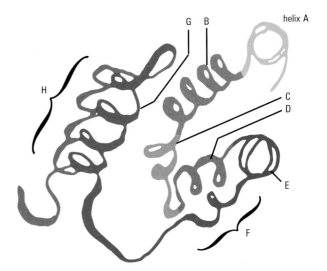

Figure 6.6 *Arrangement of coils A to H (top view)*

7. At the end of helix F, straighten out a 3 cm segment and make a broad turn back to the right. Helix G has four coils running roughly parallel to B and slightly behind and below B.

8. Stretch out a 4 cm segment to make a broad loop back to the left. Finish with five coils of helix H. There should be a short tail left over.

9. Examine the overall shape of your model. Find the cleft or open space on the left side bounded by helices C, D, E, F, and G. In the actual protein this area contains the red heme molecule that binds the oxygen.

10. Because an alpha helix is made of a stack of repeating polar peptide bonds, the helix itself is polar. The beginning of the coil is slightly positive and the end negative. As a result, helices in the same protein will spontaneously line up antiparallel to each other (that is, parallel but upside-down so that the positive end of one is next to the negative end of the other). Observe how helices A and E are antiparallel in your model, as are B and G and G and H.

11. Models of peptide backbones like this look a little empty because they do not show any R groups. In reality, myoglobin is fairly solid except for the cleft where the heme sits (figure 6.7). Place your model on a page in your report or notebook with B and G toward the paper and E and F closest to you. Trace around the perimeter of the model. The image should look like a circle with a cleft in the left side – this is more indicative of the shape of the whole protein. Given that on the scale of this model an amino acid (MW = 100 to 120) is roughly 6 to 7 mm in diameter, what size molecule would fit inside the cleft?

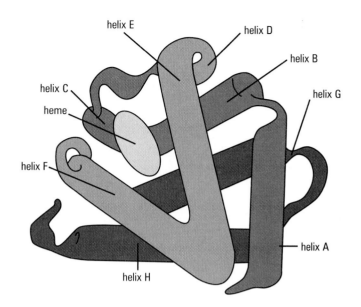

Figure 6.7   *Myoglobin*

IMPORTANT TERMS

alpha helix	electrophoresis	PAGE	SDS-PAGE
base pair (bp)	hydrogen bond	peptide	solvent front
chromatography	native conformation	R_f	
dye front	origin	SDS electrophoresis	

Laboratory Report Biological Macromolecules II: Separation, Analysis, and Modeling

FACTUAL CONTENT

Define the following terms. Brief definitions are given in the chapter, but you may also wish to consult your lecture notes and text for a better idea of how the terms are used. Pay close attention to those words that look similar but have different meanings — you will want to be able to tell them apart if they appear next to each other on a multiple-choice exam.

alpha helix

base pair (bp)

chromatography

dye front

electrophoresis

hydrogen bond

native conformation

origin

PAGE

peptide

R_f

SDS electrophoresis

SDS-PAGE

solvent front

EXERCISES - Record your data and answer the questions below.

A. DNA electrophoresis

Record what each well contains:

well	sample
1	
2	
3	
4	
5	

well	sample
6	
7	
8	
9	
10	

Record the distance between the dye spots and the sample well (the origin) after electrophoresis is complete:

Examine the pattern of stained DNA bands. Are the patterns as expected? Did the amount of nuclease you added to the DNA affect the size of the fragments?

Record the distance from each band to the origin and calculate the R_f by dividing the distance traveled by the DNA by the distance traveled by the dye front:

well number	band	distance	R_f

Graph the base-pair numbers given you by your instructor for the standards against the R_fs you observed on your gel. Draw the best straight line you can through the points.

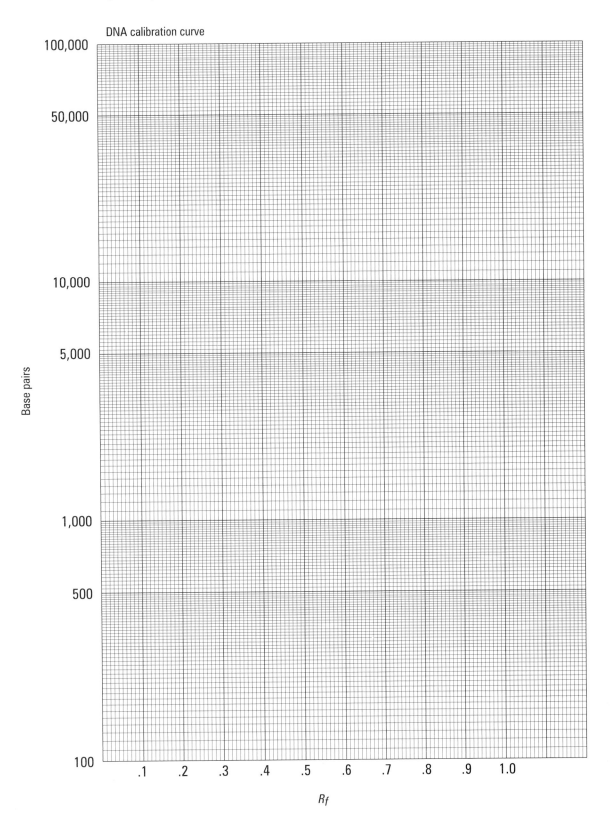

Use your calibration curve to estimate the number of base pairs in each DNA band on your gel (or, if no distinct bands are present, the high and low ends of the smeared range of DNA sizes):

B. SDS electrophoresis of proteins

Record what each well contains:

well	sample
1	
2	
3	
4	
5	
6	
7	
8	
9	
10	

Record the distance between the dye front and the origin after electrophoresis is complete:

Record the distance from each band to the origin and calculate the R_f by dividing the distance traveled by the protein by the distance traveled by the dye front:

well number	band	distance	R_f

Graph the molecular weights given you by your instructor for the standards against the R_fs you observed on your gel. Draw the best straight line you can through the points.

Protein calibration curve

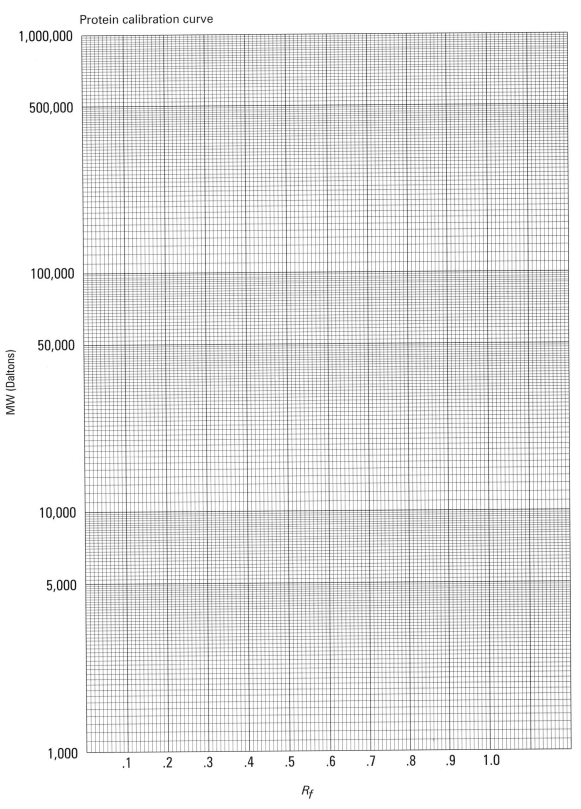

Use your calibration curve to estimate the molecular weight of each protein band on your gel:

Compare the size of your proteins to a water molecule (MW=18) and a glucose molecule (MW=180).

Examine the bands from your tissue samples. How many bands are there? How do the different tissues compare? Are there any particularly dominant bands? What size are they?

C. Modeling myoglobin structure

Place your model on the space below with B and G towards the paper and E and F closest to you. Trace around the perimeter of the model.

Given that on the scale of this model an amino acid (MW = 100 to 120) is roughly 6 to 7 mm in diameter, what size molecule would fit inside the cleft?

Enzymes

New Skills

In this lab you will use a combination of previously learned skills—making serial dilutions and standard curves and using the spectrophotometer—to assay the rate of enzyme-mediated reactions under various conditions.

phosphatase + R-O-PO$_3$ + H$_2$O —>
phosphatase + R-OH + HPO$_4^{-2}$

enzyme + substrate –>
enzyme•substrate –>
enzyme + product

Factual Content

In this chapter you will learn some basic terminology relating to enzymes in general and a few specific terms required for this particular enzyme system.

New Concepts

In doing the exercises in this chapter you will
1. observe how the presence of an inhibitor affects an enzyme.
2. observe how enzyme activity is altered by pH and temperature.
3. compare two similar enzymes to explore how behavior can vary from enzyme to enzyme.

INTRODUCTION

For the most part, lipids, carbohydrates, nucleic acids, and proteins are unique to the living world. These compounds are rarely formed by abiotic processes, and they can be extremely difficult to synthesize in even the best-equipped laboratories – the necessary chemical reactions are hard to start, and once begun, they often have to compete with side reactions that produce unwanted by-products.

It may seem odd that a mindless cell is able to do what a team of highly trained chemists cannot, but there is a good reason for this. Even the simplest cells are full of specific tools that are exquisitely formed to facilitate such difficult chemical reactions. These tools are **catalysts**, agents that make unlikely reactions more likely to occur. Cellular catalysts promote complex biological processes without encouraging side reactions.

In the last few chapters we have stressed that protein molecules are both very large and very complex. This week we will see why these two characteristics of proteins are crucial to life: they allow proteins to function as special catalysts called **enzymes**.

ENZYMES

Virtually all known enzymes are proteins (a handful are made of RNA). Every enzyme has a unique conformation that is determined by the way the protein chain folds upon itself after it is synthesized. Most enzymes have a vaguely rounded shape, but the critical feature of an enzyme's shape is a pit or cleft in its surface called the **active site**. The active site of the enzyme is lined with R groups from various amino acids of the protein that work together to create a unique chemical environment. This, coupled with the specific shape of the active site, make it an "exact fit" for some other molecule or set of molecules. The molecules that fit in an enzyme's active site are that enzyme's **substrates**.

When an enzyme and its substrate are in a solution together, the substrate molecules have a natural tendency to interact with the active sites of the enzyme molecules. Once there, however, the substrate is affected by the functional groups of the active site in such a way that the substrate molecule becomes quite unstable. The substrate changes until it is stabilized again. This change is the chemical reaction catalyzed by the enzyme. The result of the reaction is called the **product**. Product molecules do not fit the active sites as well as the substrate molecules, and they drift away from the enzyme. The enzyme is unchanged and can now attract another substrate molecule. Notice that the enzyme is specific for a particular substrate and that the substrate is converted precisely into product without error.

enzyme + substrate –> enzyme•substrate –> enzyme + product

The conformation of the enzyme is crucial to its activity. Consequently, anything that affects the conformation of a protein can affect enzyme activity. Temperature, the ionic strength and pH of a solution, and the presence of detergents and organic solvents can all influence protein conformation and thus enzyme activity. Enzymes are also affected by compounds that react with functional groups in their active sites that are necessary for catalysis. You will investigate how some of these factors influence enzyme activity in today's lab.

Because proteins are very difficult to purify, the enzymes we use in the laboratory are never 100% pure. As a result, we cannot simply measure the amount of enzyme present by mass, as some of the measurable mass is from contaminants. Instead, enzymes are measured in activity units. A formal **unit** of enzyme is defined as that *amount of enzyme* needed to convert 1 μmole of substrate to product in 1 minute in a solution of specified pH and temperature. Note that 1 unit of a relatively pure enzyme might have a mass of 1 mg, while 1 unit of the same enzyme in a less purified state may have a mass of 10 mg. Both preparations have the same amount of enzyme, but the second is weighed down with extra contaminants.

It is important to stress that units measure the amount of enzyme and not the activity that that enzyme is exhibiting at the moment. In the exercises that follow, you will always use the same amount of enzyme (about 0.1 unit). Under ideal conditions this would convert 0.1 μmoles of substrate to product per minute. However, as you change to less favorable conditions (say, an extreme temperature), this same 0.1 unit may only convert 0.02 μmoles of substrate per minute. Under these new conditions, the 0.1 unit of enzyme has an activity of 0.02 μmoles/minute.

PHOSPHATASE

The enzymes that you will be using today are called **phosphatases**. Phosphatases are common enzymes that accelerate the removal of phosphate ions from other molecules:

$$\text{phosphatase} + R\text{-}O\text{-}PO_3 + H_2O \rightarrow \text{phosphatase} + R\text{-}OH + HPO_4^{-2}$$

In this reaction R could be almost anything to which a phosphate could be bound; nucleotide monomers are one possible substrate, but there are many other possible substrates as well. There are many kinds of phosphatases: an acid phosphatase is a phosphatase that works well below pH 7, while an alkaline phosphatase would work best at high pH. In nature, acid phosphatase is used most commonly to cleave pyrophosphate, a dimer of two phosphate ions often released by other metabolic reactions, into two separate ions. It thereby "cleans up" a metabolic waste by turning it into usable phosphate. (The human prostate gland is rich in acid phosphatase, and the assay you will use today is used clinically to help diagnose prostate cancer.) Alkaline phosphatases catalyze the same reactions but occur in basic fluids and secretions.

Phosphatase happens to be a good example of how enzymes can be regulated by **inhibition**. In this case the enzyme is inactive in the presence of excess phosphate ions. Thus, if the cell is already full of phosphate, the phosphatase enzymes will automatically shut down and not release any more phosphate. Different enzymes are inhibited by different things, not all of which serve a natural regulatory function. Many drugs are artificially produced enzyme inhibitors.

EXERCISES WITH ENZYMES

Reagents

nitrophenol standard	0.24 mM nitrophenol
sodium hydroxide solution	0.2 N NaOH
acid phosphatase solution	0.2 units/ml in 0.01 M Na acetate, pH 5.0
alkaline phosphatase solution	0.2 units/ml in 0.01 M THAM, pH 9.0
assay buffers	0.1 M acetate; 0.1 M glycine; 0.1 M THAM; pH 3.0, 5.0, 6.5, 7.5, and 9.0
substrate solution	7 mM p-nitrophenyl phosphate
phosphate solution	0.5 M Na_2HPO_4

A.   Calibrating phosphatase activity

Measuring the amount of product produced by an enzyme can be difficult: the reaction must be precisely timed, the product identified and quantitated. The assay you will use today is simpler than most, as it uses a spectrophotometer to directly measure product concentration after the reaction has been stopped. This is not possible for many enzymes, since few products are colored, but it happens that phosphatase will turn the artificial substrate **nitrophenyl phosphate** into a product, **nitrophenol**, which is yellow in alkaline solutions. For this assay the reaction will be stopped after five minutes by adding a quantity of strong base, which simultaneously inactivates the enzyme – the base is strong enough to denature the enzyme – and develops the color of the product. The entire solution can then be measured in the spectrophotometer and compared to a standard curve to calculate the actual product concentration. Even though this assay requires no specialized chemical techniques, it is vital that you measure your reagents as carefully as possible in order to obtain reproducible results.

1.   Turn on the spectrophotometer and adjust it as directed in chapter 5. Set the wavelength to 410 nm.

2.   Obtain 5 ml of the nitrophenol standard. This has a concentration of 0.24 mmoles of nitrophenol per liter, which means that it has 0.24 μmol of nitrophenol per ml. Put exactly 2.5 ml in each of two test tubes. Label the first "0.60 μmoles" (that is, 2.5 ml x 0.24 μmol per ml). Dilute the second with 2.5 ml of water and label it "0.30 μmoles." Take 2.5 ml of the second solution and dilute it to 0.15 μmoles and continue making serial dilutions until you have tubes with 0.60, 0.30, 0.15, 0.075, and 0.038 μmoles of nitrophenol. Each tube should have 2.5 ml (discard any leftover solution). Set up a final tube with 2.5 ml of water to represent 0 μmoles of nitrophenol.

3.   Add 5 ml of 0.2 M NaOH solution to each tube. They should turn varying shades of yellow.

4.   Transfer the solutions to spectrophotometer tubes and, using the 0 μmole tube as a blank, measure the absorbance of each solution in the spectrophotometer. Graph the absorbance versus the amount of nitrophenol and draw the best straight line you can through the points. This will be your standard curve for all of the following assays.

B.   Determining standard phosphatase activity

There are thousands of known enzymes and each one is different. It is impossible for us to examine them all, but we can observe some important principles simply by comparing two similar enzymes. For this reason half the class will use acid phosphatase for these exercises and half will use alkaline phosphatase. The procedure is exactly the same except that for acid phosphatase the standard buffer will be at pH 5 and for alkaline phosphatase it will be pH 9.

1.   Put 1.5 ml of the appropriate standard buffer into each of two clean test tubes. Add 0.5 ml water to one tube (as a negative control) and 0.5 ml of enzyme to the other. Make sure the solutions are at room temperature before proceeding (i.e., if one of the stock solutions was cold, let the mixtures stand at room temperature 5 minutes).

2.   Add 0.5 ml of substrate to each tube. Begin timing the reaction as soon as the substrate is added to the enzyme tube.

3.   After 5 minutes, add 5 ml of NaOH solution to each tube to stop the reaction.

4.   Transfer the solutions to spectrophotometer tubes and read the absorbance of each, using the same blank as in part A.

5.   Use your standard curve to calculate the amount of nitrophenol present in each tube. The control will usually be very close to the blank, indicating about 0 μmoles of nitrophenol present (sometimes the control will be higher than 0, especially at high temperatures in exercise D).

6.   Subtract the amount of product in the control tube from the amount in the enzyme tube to determine how much product was produced by the enzyme. Divide this number by 5 to calculate how much was produced each minute. Since these are the optimum conditions, the number of μmoles per minute is the same as the number of units of enzyme present. If it is not close to 0.1 unit, you will have to adjust the amount of enzyme you use in the other experiments accordingly.

C.   Inhibition by phosphate

Repeat exercise B, except add 0.1 ml of phosphate solution to each tube. Record the activity observed.

Because of time considerations, each group will only do one of the following exercises.

Before you begin, hypothesize which temperature or pH you will be testing will give you the highest activity and which will give you the lowest activity. After you are done, present your results to the class so everyone can observe the effect of that variable on your enzyme.

D.   Effect of temperature

Investigate the effects of temperature by performing the assay at 0°C (ice bath), 40°C (controlled water bath or incubator), 60°C (water bath), 80°C (water bath), and 100°C (boiling water bath), all at standard pH. Be sure to equilibrate the buffer/enzyme mixture to the experimental temperature for 5 minutes before adding the substrate. Graph the activity versus temperature (you can use exercise B to give the activity at room temperature, about 20°C). Was your hypothesis confirmed by the experiment?

E.   Effect of pH

Investigate the effects of pH by performing the assay at pH 3.0, 5.0, 6.5, 7.5, and 9.0 using the other assay buffers. These should be done at room temperature. Graph the activity versus pH (you have already done one of these pH values as exercise B). Was your hypothesis confirmed by the experiment?

F.   Comparison of the two enzymes

Each group will present their data to the whole class. How do the two enzymes compare

1.   In their reaction to the presence of phosphate?

2.   With differing pH? What causes the change in activity as pH changes? Does this suggest anything about the types of R groups present in each enzyme?

3.   With differing temperature? Can you explain why the activity is affected by temperature in the way that it is?

IMPORTANT TERMS

active site	nitrophenol	substrate
catalyst	nitrophenyl phosphate	unit
enzyme	phosphatase	
inhibition	product	

Laboratory Report Enzymes

FACTUAL CONTENT

Define the following terms. Brief definitions are given in the chapter, but you may also wish to consult your lecture notes and text for a better idea of how the terms are used. Pay close attention to those words that look similar but have different meanings – you will want to be able to tell them apart if they appear next to each other on a multiple-choice exam.

active site

nitrophenyl phosphate

catalyst

phosphatase

enzyme

product

inhibition

substrate

nitrophenol

unit

EXERCISES - Record your data and answer the questions below.

A. Calibrating phosphatase activity

Complete the table with the measured absorbance of each dilution of nitrophenol:

amount of nitrophenol	absorbance at 410 nm
0 μmoles	
0.038 μmoles	
0.075 μmoles	
0.15 μmoles	
0.30 μmoles	
0.60 μmoles	

In the space provided graph the absorbance versus the amount of nitrophenol and draw a line through the points.

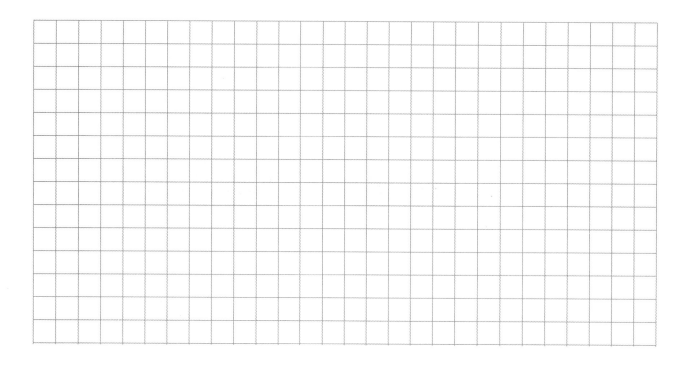

B. Determining standard phosphatase activity

You will be using either acid phosphatase or alkaline phosphatase for all of these exercises. Record which enzyme you are using and the pH of the standard buffer for that enzyme:

Complete the table with the measured absorbance for each tube and the corresponding amount of nitrophenol, as indicated by your calibration curve:

tube	absorbance at 410 nm	amount of nitrophenol
enzyme		
control		

Subtract the amount of product in the control tube from the amount in the enzyme tube to determine how much product was produced by the enzyme:

Divide this number by 5 to calculate how many μmoles of nitrophenol was produced by the enzyme each minute:

How many units of enzyme were present in the enzyme tube?

Obtain the results for this exercise from a group using the other enzyme and record the number of units of enzyme they measured along with the pH of their standard buffer:

C. Inhibition by phosphate

Complete the table with the measured absorbance for each tube and the corresponding amount of nitrophenol, as indicated by your calibration curve:

tube	absorbance at 410 nm	amount of nitrophenol
enzyme + phosphate		
control		

Subtract the amount of product in the control tube from the amount in the enzyme tube to determine how much product was produced by the enzyme:

Divide this number by 5 to calculate how many μmoles of nitrophenol was produced by the enzyme each minute:

What is the effect of phosphate on enzyme activity?

Obtain the results for this exercise from a group using the other enzyme and record the enzyme activity they measured in the presence of phosphate:

How do the two enzymes compare in their reaction to the presence of phosphate?

What would happen to the concentration of phosphate inside a cell as these enzymes work? What would that changing phosphate concentration do to the enzymes' activity? How might this effect be beneficial to the cell?

D and E. Effect of temperature and pH – raw data

You will be testing just one enzyme (acid or alkaline phosphatase) and one variable (temperature or pH) in this exercise. Use the following table to record your data, and then combine that data with the data obtained by other groups in the class to complete the tables in the following sections.

Record which enzyme you are using and which variable you are testing:

Complete the table with the measured absorbance for each tube and the corresponding amount of nitrophenol, as indicated by your calibration curve:

value of variable (temperature or pH)	tube	absorbance at 410 nm	amount of nitrophenol
	enzyme control		
	enzyme control		
	enzyme control		
	enzyme control		
	enzyme control		

Subtract the amount of product in the control tube from the amount in the enzyme tube and divide this number by 5 to calculate the activity of the enzyme for each value of the variable. Record these activities in the appropriate table below.

D. Effect of temperature

Based on what you know about hydrolysis reactions (chapter 4, exercises D and E) and protein denaturation (chapter 5, exercise C), form a hypothesis as to how temperature will effect the activity of the enzymes:

Complete the tables based on the experiments performed by the class:

acid phosphatase, pH 5

temperature	activity (μmoles/min)
0°C	
20°C	
40°C	
60°C	
80°C	
100°C	

alkaline phosphatase, pH 9

temperature	activity (μmoles/min)
0°C	
20°C	
40°C	
60°C	
80°C	
100°C	

In the space provided, graph the activity of each enzyme versus temperature and draw a curve through each set of points.

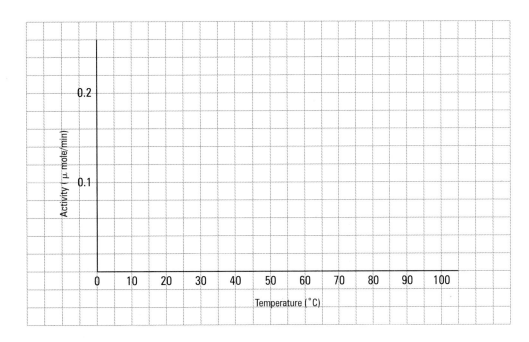

How do the temperature profiles for each enzyme compare?

Was your hypothesis confirmed by the experiment? If not, can you modify your hypothesis to fit the new data?

E. Effect of pH

Based on what you know about these enzymes (exercise B from this chapter) and protein denaturation in general (chapter 5, exercise C), form a hypothesis as to how pH will effect the activity of each enzyme:

Complete the tables based on the experiments performed by the class:

acid phosphatase, 20°C

pH	activity (μmoles/min)
3.0	
5.0	
6.5	
7.5	
9.0	

alkaline phosphatase, 20°C

pH	activity (μmoles/min)
3.0	
5.0	
6.5	
7.5	
9.0	

In the space provided, graph the activity of each enzyme versus pH and draw a curve through each set of points.

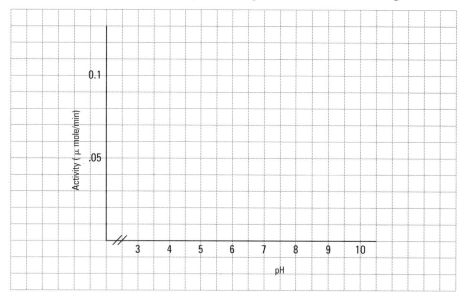

How do the pH profiles for each enzyme compare?

Was your hypothesis confirmed by the experiment? If not, can you modify your hypothesis to fit the new data?

Cellular Respiration

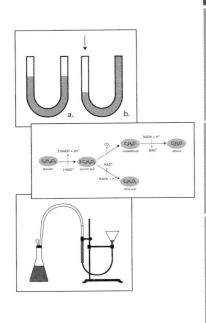

OBJECTIVES

New Skills

After working on the exercises in this chapter, you will learn how to
1. use aseptic technique to transfer bacteria to sterile test media so that the ability of the bacteria to use fermentation can be studied.
2. collect CO_2 produced during fermentation under constant pressure so that it can be accurately measured as an indication of fermentation rate.

Factual Content

In this chapter you will learn some basic terminology relating to respiratory and energy-releasing reactions.

New Concepts

In doing the exercises in this chapter, you will
1. compare the fermentation products made by various organisms.
2. compare the rate of fermentation by yeast using different carbohydrates as food.
3. observe how mitochondrial enzymes can oxidize organic molecules and use the energy so released to reduce other compounds.
4. observe how a mixture of a few chemicals can organize itself into a complicated chemical cycle.

INTRODUCTION

When an enzyme catalyzes a reaction, it merely speeds up a reaction that sooner or later would occur on its own – the enzyme does not have an internal energy source with which it can force a reaction to occur. (This is admittedly a fine distinction, as some enzyme-mediated reactions are so slow without an enzyme that they might as well not be occurring at all.) In nature, a chemical reaction takes place when it converts a molecule that contains a great deal of energy into a molecule that contains less energy, releasing the energy difference to the environment. Nature always favors converting energy-rich compounds into energy-poor compounds.

The problem is that cells depend on energy-rich compounds to survive: it takes a lot of energy to weld monomers into polymers, for example. Cells manufacture energy-rich compounds in complex, step-by-step processes, in which key steps are linked to energy-releasing reactions. That way, every time the cell puts some energy into a chemical bond, it releases even more energy from a different bond so that there is still a net release of energy.

The reaction most often used by cells to release energy to power other reactions is the removal of one or two phosphates from the nucleotide **adenosine triphosphate (ATP)**. This reaction produces small bursts of energy sufficient to power individual steps in metabolic pathways. Of course, sooner or later the ATP will run out. It can, however, be regenerated if an even more energetic molecule is sacrificed. Monosaccharides, especially glucose, are perfect sources of this additional energy.

Theoretically, a molecule of glucose contains enough energy to make about 100 molecules of ATP, so making ATP can be thought of as "making change," like cashing a $1 bill into pennies. Unfortunately, this process wastes a lot of energy, so that even the most efficient organisms are unable to generate more than 38 ATPs from a molecule of glucose. (This may not seem very efficient, but man-made machines, by comparison, usually waste at least 75% of the energy they consume.) The release of this amount of energy from glucose requires oxygen-driven redox reactions. Because of this oxygen requirement, the process by which a cell releases energy by breaking down glucose into carbon dioxide and water is called **cellular respiration**, even though individual cells do not "breathe" in the same sense that animals do.

OXYGEN UTILIZATION AND FERMENTATION

All cells can metabolize glucose and transfer the energy to ATP. The entire process of glucose metabolism includes several distinct enzymatic pathways. The initial process is called **glycolysis** and results in splitting the glucose into two 3-carbon molecules of pyruvic acid. Glycolysis does not do anything with this pyruvic acid, and the energy released by this process is small – enough to make just two molecules of ATP.

After glycolysis, a cell must use some other pathway to metabolize the pyruvic acid. Several such pathways exist, and which one a cell uses depends on what kind of cell it is and whether or not the cell is able to use oxygen. If the cell can use oxygen, it can ultimately "burn" the pyruvate molecules completely to carbon dioxide and water and generate about 34 more ATPs (some bacteria are efficient enough to obtain 36 ATPs this way). Oxygen-rich environments, as well as organisms and metabolic processes that require oxygen, are all described as **aerobic**. (Thus cellular respiration is often referred to as aerobic respiration.) Environments and processes that do not involve oxygen are called **anaerobic**.

Anaerobic organisms can live on the small amounts of energy released by glycolysis. They use the process of **fermentation** to convert pyruvic acid into some other kind of organic compound that is then excreted as a waste product. There are actually many kinds of fermentation, each method converting the pyruvic acid into a different product.

Fermentation is most common in species of anaerobic bacteria. Fermenting bacteria usually produce some kind of organic acid as their waste. Many also release a gas, such as CO_2 or hydrogen, in addition to the organic product.

Fermentation is of considerable economic importance. Many industrial chemicals are produced by fermentation. Bacterial fermentation is also important in the food industry. For example, the yogurt you looked at in chapter 3 was formed when fermenting bacteria produced enough acid to curdle the milk they were grown in. Bacteria produce cheese, too – the various kinds of cheese are made by different species of bacteria and flavored by different organic acids. If cheesemakers use a gas-forming bacterium, then their cheese will have bubbles in it (like Swiss cheese). Fermenting bacteria were discovered by Louis Pasteur,

and heating milk to kill bacteria and keep it from going sour is called **pasteurization**.

FERMENTATION IN EUKARYOTES

Eukaryotic cells usually depend on aerobic respiration, not fermentation, although some cells can survive for short periods on fermentation if they have to. For example, if you exercise vigorously, your muscle cells may run out of oxygen and begin to use a fermentation process that produces **lactic acid**. As the lactic acid builds up, your muscles become fatigued, until you are forced to rest and allow your body to return to normal. Likewise, plant cells can switch to fermentation, which in their case produces **ethanol** (sometimes called grain alcohol) and carbon dioxide. Plant roots are exposed to low oxygen conditions whenever the soil they live in becomes saturated with water, filling in the tiny air spaces normally present in the soil. In these situations fermentation is vital if the roots are to survive.

One group of eukaryotes that can survive indefinitely on fermentation includes the unicellular fungi called yeasts. Like plants, yeasts excrete ethanol and CO_2 as fermentation products. Virtually all alcoholic beverages are made by yeast fermentation, and the bubbles of CO_2 that yeasts generate inside bread dough are what make bread rise. Baker's yeast is one of the easiest organisms to use in the laboratory when studying fermentation.

THE CHEMISTRY OF FERMENTATION

Fermentation is complicated. Why don't cells simply excrete pyruvic acid as a waste product instead of converting it into something else first?

When a cell is unable to use oxygen, an atomic traffic jam occurs. This is because two hydrogen atoms must be removed from glucose before it can give up its energy to ATP. The removal of hydrogen is a combination of an acid-base reaction (loss of H^+) and an oxidation reaction (loss of an electron). Since the acid-base aspect of dehydrogenation is mitigated by the cell's buffering system, *we can treat any loss of hydrogen atoms as oxidation and any gain of hydrogen atoms as reduction*. (This is a fundamental concept in biochemistry.) The hydrogens are temporarily stored on two molecules of the nucleotide dimer **NAD**, which is thus turned into **NADH**. Removing hydrogen from NADH is not a problem if oxygen is available, as oxygen can react with the hydrogen to form water, which is harmless and chemically stable (this means that the glucose is ultimately oxidized by oxygen). Without oxygen, the NAD can be completely converted into NADH, which means that there will be nowhere left to put the hydrogens being taken from the glucose. In this event the system shuts down.

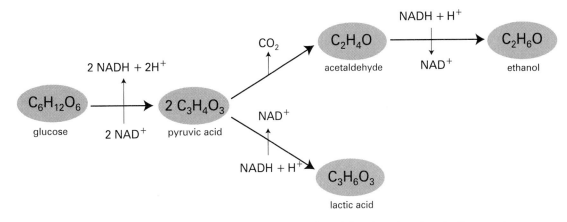

Figure 8.1 *Fermentation reactions in plants and fungi (top) and animals (bottom)*

Fermentation is the process of regenerating NAD by removing hydrogen from NADH and combining it with pyruvic acid. Fermentation does not release any more energy, but by recycling the cell's NAD it guarantees that glycolysis can continue. Animal cells form lactic acid by attaching the hydrogens directly to pyruvic acid molecules. In plants and fungi the pyruvic acid is first converted into **acetaldehyde** and CO_2. The acetaldehyde is then reduced by NADH to produce ethanol (figure 8.1). As you will see in today's lab, it is possible to inhibit this kind of fermentation with **sodium bisulfite**, a chemical that reacts with acetaldehyde before it can be reduced to ethanol.

THE KREBS CYCLE AND THE ELECTRON TRANSPORT CHAIN

Glycolysis followed by fermentation liberates only 2% of the energy in a glucose molecule in usable form – the rest is trapped in waste products such as lactic acid or ethanol. When we say that a molecule like ethanol contains energy, we mean that under some conditions it becomes unstable and will react in such a way as to release energy. We know that this is the case with ethanol because it will burn. Ideally then, a cell should try to transfer the hydrogen atoms taken from glucose to the most stable molecules it can – the stable molecules would have very little energy left and the remainder would be released for the cell to use. The most stable compounds that can be formed from the elements present in carbohydrates are carbon dioxide and water. Neither will burn, and, in fact, energy must be pumped into these molecules before they will come apart. Glucose can be converted into carbon dioxide and water by the addition of oxygen, and this is the principle behind aerobic respiration. The process is more complex than glycolysis, requiring many groups of enzymes working together. In eukaryotic cells these enzymes are all packaged together in the mitochondria.

Pyruvic acid enters the mitochondria and begins a series of reactions called the **Krebs cycle**. (In aerobic respiration, NADH is converted back to NAD by the electron transport chain, described later.) The Krebs cycle binds the pyruvate to a carrier and slowly breaks it apart, releasing carbon dioxide and hydrogen. At this point the hydrogen is mainly held by more NAD, and little usable energy is produced. Thus the Krebs cycle mainly serves to oxidize the carbon backbone of the pyruvate into CO_2 and free up hydrogen atoms.

The inner mitochondrial membrane contains several enzymes that work together as the **electron transport chain**. NADH enters one end of the chain and gives up its hydrogen, which is transported across the membrane to react with oxygen and form water. (This is called the electron transport chain because the energy from reduction is carried only by the electrons.) Because NADH is a higher-energy compound than water, energy is released during transport. The transport chain is coupled to ATP-regenerating enzymes that can use this energy. The electron transport chain ends the process begun by the Krebs cycle by generating water and ATP.

SELF-ORGANIZING CYCLIC REACTIONS

In chapter 7 you looked at a simple reaction catalyzed by a single enzyme – there was one substrate, one product, and a constant reaction rate for any particular set of conditions. In exercise C you will see a similar reaction catalyzed by an enzyme isolated from mitochondria. However, cellular respiration as a whole is much more complicated than this. It involves dozens of reactions, each with its own enzyme. The product of one reaction may be a potential substrate for several other reactions. It might seem like such a complex web of reactions would be difficult to maintain without some kind of central control, but, amazingly, this is not the case. In some cases chemical systems are able to organize themselves into complex cycles. The key to keeping such a system going is how it releases energy.

If exposed to a hot enough flame, glucose and other sugars will catch fire and burn. Obviously, such a rapid release of energy would be damaging to the cell, which is why the cell's oxidative reactions are constrained to proceed at a more orderly, stepwise pace. When a relatively large amount of energy is released in a slow, limited fashion, unusual things can occur as the intermediate states develop and interact with each other. Technically, such systems are described as being *nonlinear*, because there is no simple way for the high-energy compounds to quickly release their energy. Simple mathematical models are called *linear* (that is, when graphed they produce a straight line).

An analogous situation sometimes arises in the weather when a mass of warm air contacts a mass of cold air. The simplest result for such a system would be for the two masses to mix and form a single mass of air that is intermediate in temperature, but that is not always possible. When it is not possible, the weather system may spontaneously organize itself into a complex structure like a tornado, which can then dissipate the energy difference in a new and unexpected way. Over the past 30 years chemists and physicists have become very interested in **nonlinear dissipative systems** because of their abilities to organize themselves into complex structures.

Living systems in general, and the Krebs cycle in particular, are nonlinear dissipative systems. They behave differently from simple chemical reactions, which is why they can form stable cycles. In the 1960s two Russians, Belousov and Zhabotinsky, happened across another self-organizing chemical system while studying the Krebs cycle. This so-called **BZ reaction** is not exactly the same as the Krebs cycle, because it uses no enzymes and the ultimate oxidant in it is not oxygen but bromine. However, it is similar in that it involves the stepwise oxidation of an organic acid (we will use malonic acid) to CO_2. It also uses a catalyst that contains iron atoms, just as many Krebs cycle enzymes do. What is fascinating about this reaction is that, after mixing together a few simple chemicals, the reaction organizes itself into a complicated cycle of several intermediate reactions that periodically interact with the iron catalyst to change its color from blue to red and back again. Because of this color oscillation, the reaction is sometimes called a **chemical clock**. When spread out in a shallow container, the oscillating colors create intricate patterns of colors that move in waves across the dish. Biologists are interested in chemical clocks not only as models of the Krebs cycle, but because their self-organizing behavior suggests how complex life may have first arisen from a mixture of nonliving chemicals. In exercise D you will set up the BZ reaction and observe for yourself how the cyclic behavior organizes itself as the reaction progresses. This is as close as you can come to observing the full complexity of the Krebs cycle in the laboratory.

EXERCISES WITH CELLULAR RESPIRATION

A. Testing bacteria for fermentation

Not all bacteria can use fermentation. Those that can use a variety of pathways that lead to a variety of products. The following procedure is commonly used by bacteriologists to test whether a species of bacteria can ferment and to get a rough idea of what fermentation products the bacteria can make.

Most bacteria that ferment produce some kind of organic acid. These acids can be easily detected by acid-base indicators such as phenol red. Phenol red is red in neutral or basic solutions but turns yellow in the presence of acid. Therefore, fermentation will make a tube of phenol red broth turn yellow. Bacteriologists usually place a small upside-down tube (called a Durham tube) inside test tubes of phenol red broth. That way, if gas is released along with the acid, it will be trapped inside the Durham tube and be visible as a bubble.

There are three possible outcomes for this test: no acid-forming fermentation, production of acid, or production of acid and gas. We will illustrate this with three positive controls: *Bacillus megaterium*, a nonfermenter; *Enterococcus faecalis*, an acid producer; and *Escherichia coli*, an acid/gas producer. Each pair of students will prepare a culture tube for one of these controls. You will also prepare a tube of *Staphylococcus epidermidis* as an unknown. By comparing the results from the unknown with the controls, you should be able to determine whether *Staphylococcus epidermidis* can ferment, and if so, whether it produces both acid and gas or just acid.

Bacteria are everywhere. Watch closely as your instructor demonstrates how to use sterile techniques to prevent unwanted bacteria from contaminating a culture. Use these techniques in your own work or it will not be reproducible.

Reagents

phenol red broth	bacterial growth medium containing phenol red indicator and glucose

bacterial cultures	pure cultures of *Bacillus megaterium*, *Escherichia coli*, *Enterococcus faecalis,* and *Staphylococcus epidermidis*

PERIOD ONE

1. Flame an inoculating loop to red-hot. As the loop cools, remove the caps from a sterile phenol red tube and one of the culture tubes of a control organism. Pass the mouths of both tubes through the flame. Dip the loop into the broth culture and then inoculate the phenol red. Flame both tubes again and replace the caps. Flame your loop when you are done. Prepare one or more tubes as your instructor assigns.

2. Repeat the procedure using an inoculum of *Staphylococcus epidermidis*.

PERIOD TWO

1. Examine the tubes for signs of fermentation. How much gas was produced by *Escherichia coli*? Did *Staphylococcus epidermidis* ferment? If so, did it produce any gas?

B. Fermentation in yeast

A yeast cell releases two molecules of carbon dioxide for every molecule of glucose it ferments. Therefore, one way to study fermentation is to monitor the release of CO_2. It may seem like a simple matter to measure the amount of CO_2 generated by fermentation, but the problem is more complicated than it first appears.

Gas must, of course, be collected in a sealed vessel or it will escape. However, the amount of gas that will fit into a sealed vessel varies depending on the pressure of the gas. Imagine, for example, a cylinder sealed at the top by a movable piston (figure 8.2a). If pressure is applied to the piston, it will move into the cylinder and reduce the volume of the gas inside (figure 8.2b). The same amount of gas is still inside the cylinder (because it is sealed), but the pressure from the piston has compressed it into a smaller volume. Simply measuring the volume of the container does not tell us how much gas is present unless we can also account for the pressure the gas is under.

A sealed bottle with a culture of fermenting yeast will continue to produce CO_2 until the yeast runs out of sugar to ferment. The CO_2 will collect inside the bottle under increasing pressure because the volume of the bottle cannot change (this is how the bubbles get into fermented beverages like beer and champagne and why champagne can be under so much pressure before it is opened). If we want to get an accurate measurement of how much CO_2 is actually produced, we must collect it in a container that automatically expands so that the pressure inside stays the same as the air pressure outside.

One way to do this is to seal one end of the collection vessel with a tube of water. Variations in pressure will cause the water to move in the tube and allow us to monitor any pressure buildup. How this works is illustrated by figure 8.3. If water is placed inside a U-shaped tube open on both sides, it will fill both sides to the same level (8.3a). As long as the air pressure is the same on both sides of the tube, the water will remain level. However, if pressure is applied to one side, the water will rise in the opposite side (8.3b), alerting us to a pressure imbalance (this is how most barometers measure air pressure).

The complicated apparatus shown in figure 8.4 uses this principle to measure the CO_2 generated by fermentation. In this apparatus, a flask containing yeast and sugar is connected by a tube to a measuring pipet. The bottom of the pipet is connected to a funnel of water, which finds its level in the pipet. As gas enters the pipet from the fermentation, it builds up pressure and pushes the water back into the funnel. The pressure buildup can be relieved in the pipet by sliding it upwards until the water level inside is again level with the water in the funnel. At that point the volume of gas at normal air pressure can be read.

Fermentation follows directly from glycolysis under anaerobic conditions. This means that the ultimate fuel for fermentation is glucose. Most cells can convert other carbohydrates into glucose, however, so alternate fuels are possible. The class will try several carbohydrates in this experiment.

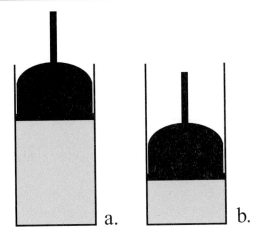

Figure 8.2 *Relation of gas volume to pressure*

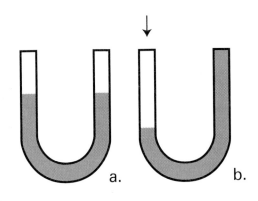

Figure 8.3 *Effect of pressure on water level*

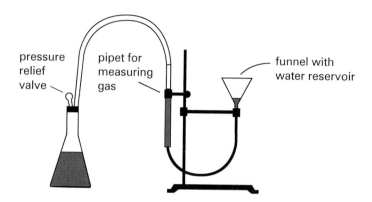

Figure 8.4 *An apparatus for measuring CO$_2$ produced during fermentation*

Reagents

buffer	0.02 M KH$_2$PO$_4$, pH 5
yeast culture	5 g baker's yeast per 200 ml H$_2$O
carbohydrate solutions	0.1 M glucose, fructose, lactose, or sucrose
inhibition solution	0.1 M glucose; 1% Na bisulfite

1. In a 125 ml flask, mix 50 ml of buffer, 25 ml of yeast culture, and 50 ml of either one of the carbohydrate solutions or the inhibition solution. Your lab instructor will assign different carbohydrates to different groups; one group will use water instead of carbohydrate as a control. In your notes, write a hypothesis predicting whether you will see much CO$_2$ production with your assigned solution.

2. Assemble the apparatus as shown in figure 8.4. Keep the pressure relief valve open while you set up so that any gas given off before the experiment starts does not build up.

3. Adjust the pipet up or down in its holder until the 0 line is just level with the bottom of the holder.

4. Add or remove water from the funnel until the water level in the pipet is at the 0 mark. (Since the valve is open, both sides should be at room pressure and the water levels should be equal.)

5. Gently swirl the flask to release any bubbles, and close the valve to begin the experiment. Gas should slowly start to build up in the pipet. This will push the water level down in the pipet.

6. After 10 minutes, gently swirl the flask again to release any bubbles. Slide the pipet up until the water level returns to its original position at the bottom of the pipet holder.

7. The number now at the water level shows how much gas (in ml) has collected. Record this in your notes.

8. Repeat steps 6 and 7 every 10 minutes for an hour or until 0.9 ml of gas are collected.

9. Graph your results. Was your hypothesis confirmed? Compare your results with those obtained by other groups. Does the rate of fermentation depend on the carbohydrate fermented? Is there a difference between disaccharides and monosaccharides? What is the effect of bisulfite on CO$_2$ production? Why?

C. Succinic dehydrogenase

Like glycolysis, the Krebs cycle is a series of reactions with each step being catalyzed by its own enzyme. For this exercise we will focus on a single enzyme of the Krebs cycle: **succinic dehydrogenase**. Succinic dehydrogenase is a complex enzyme that has several chemical groups bound to its protein backbone, including eight iron atoms. These extra groups are critical in allowing the enzyme to catalyze its redox reaction, the removal of a pair of hydrogens from succinic acid, an intermediate product in the Krebs cycle. Normally the hydrogens would be passed on to **FAD**, a molecule similar to NAD, but in this procedure you will use the hydrogen to reduce the blue dye dichlorophenolindolphenol. As the reaction proceeds, the dye should become decolorized by reduction.

Structurally, succinic acid has two carboxyl groups separated by two hydrocarbon groups. A hydrogen is taken from each of the hydrocarbon groups so that the central carbons double bond to form fumaric acid:

$$HOOC-CH_2-CH_2-COOH \longrightarrow HOOC-CH=CH-COOH + 2H$$

The enzyme can be "fooled" into accepting another compound, malonic acid, as a substrate. Malonic acid also has two carboxyl groups but only one hydrocarbon in the middle, so it cannot form a double bond:

$$HOOC-CH_2-COOH$$

You will not use a purified enzyme for this investigation, but a crude cell fractionation. That means that many side reactions will occur, but the presence of large amounts of substrate should make the reaction catalyzed by succinic dehydrogenase noticeable above the rest. The fractions will be separated by centrifugation, and you will have a chance to find where in the cell the enzyme is located. You will test both succinic acid and malonic acid to see if they can be oxidized and used to reduce (decolor) a dye.

The dye used in this procedure is sensitive to many conditions, and not all of the color changes you observe will be due to enzymatic reduction. This happens to be a very good example of the importance of proper controls. By comparing the observed color changes with the different reagent combinations, you should be able to figure out what happens in each tube. Although this investigation requires no calculations, it does turn out to be an interesting logic problem.

Reagents

KCl solution	1.14% KCl
buffer	0.1 M phosphate, pH 7.4
dye solution	0.01% dichlorophenolindolphenol
succinate solution	0.1 M sodium succinate
malonate solution	0.1 M sodium malonate
janus green stain	0.2% janus green B

1. Your lab instructor will begin by homogenizing about 20 g of tissue (either fresh wheat germ or beef liver) in a blender with 100 ml of cold KCl solution. The homogenate will be filtered through cheesecloth and then separated by centrifugation. The supernatant will be removed and saved; the pellet will be resuspended in 200 ml of cold KCl solution. Both should be kept on ice until use.

2. Before you set up the reaction, put a drop of each suspension on separate microscope slides and stain with janus green (cf. chapter 3). Can you find mitochondria? Which preparation has the most? Use this observation to hypothesize which suspension contains the most succinic dehydrogenase. Record this hypothesis in your notes – it will be tested in step 4.

3. In each of six test tubes mix 0.5 ml of buffer and 0.1 ml of dye.

4. Add to the test tubes the following solutions:

   A. 0.5 ml supernatant + 0.25 ml succinate
   B. 0.5 ml supernatant + 0.25 ml water
   C. 0.5 ml pellet + 0.25 ml succinate
   D. 0.5 ml pellet + 0.25 ml water
   E. 0.5 ml supernatant + 0.25 ml malonate
   F. 0.5 ml water + 0.25 ml succinate

5. Incubate the tubes at 37°C for 30 minutes.

6. Compare the color of the different tubes. In which tubes was the dye reduced? Why wasn't it reduced in the other tubes?

D. The BZ reaction

Again, we must emphasize that the BZ reaction bears only a superficial resemblance to the Krebs cycle – they are not the same at all. It is much easier to work with than the entire Krebs cycle, however. In the following activity you will simply mix the indicated chemicals and watch the colors and patterns that develop. The patterns that develop indicate that cyclic reactions have established themselves (this has been verified with sophisticated redox measurements that we will not attempt here). If the Krebs cycle produced colored intermediate compounds, it might look similar to this chemical clock.

Reagents

potassium bromate solution	7% KBrO$_3$; 3 % H$_2$SO$_4$
malonic acid solution	10% malonic acid
NP-40 solution	0.1% NP-40
ferroin indicator	0.025 M phenanthroline ferrous sulfate

1. Measure 6 ml of potassium bromate solution in a 10 ml graduated cylinder. Add 1 ml of malonic acid solution and 1 to 2 drops of NP-40 solution. (NP-40 is a mild detergent that helps the solution spread out in a thin film in the petri dish.)

2. Place 1 ml of ferroin indicator in the center of a plastic petri dish. It should be blood red in color.

3. Add the bromate/malonate mixture from step 1 to the ferroin indicator and swirl the petri dish to mix. The color should change to light blue.

4. Observe the reaction. Does a color cycle establish itself? Wait a few minutes – do you see evidence of the cycles segregating themselves in different microenvironments? Check the reaction again after 10 or 15 minutes. Do you see gas bubbles? What kind of gas might this be?

IMPORTANT TERMS

acetaldehyde	cellular respiration	fermentation	NADH
adenosine triphosphate (ATP)	chemical clock	glycolysis	nonlinear dissipative system
aerobic	electron transport chain	Krebs cycle	pasteurization
anaerobic	ethanol	lactic acid	sodium bisulfite
BZ reaction	FAD	NAD	succinic dehydrogenase

Laboratory Report Cellular Respiration

FACTUAL CONTENT

Define the following terms. Brief definitions are given in the chapter, but you may also wish to consult your lecture notes and text for a better idea of how the terms are used. Pay close attention to those words that look similar but have different meanings – you will want to be able to tell them apart if they appear next to each other on a multiple-choice exam.

acetaldehyde

adenosine triphosphate (ATP)

aerobic

anaerobic

BZ reaction

cellular respiration

chemical clock

electron transport chain

ethanol

FAD

fermentation

glycolysis

Krebs cycle

lactic acid

NAD

NADH

nonlinear dissipative system

pasteurization

sodium bisulfite

succinic dehydrogenase

EXERCISES - Record your data and answer the questions below.

A. Testing bacteria for fermentation

Complete the table to indicate the appearance of the phenol red tubes before and after bacterial growth:

sample	reaction	color of medium	gas in Durham tube
before inoculation	no fermentation	red	none
Bacillus megaterium	non-fermenter		
Enterococcus faecalis	acid producer		
Escherichia coli	acid/gas producer		
Staphylococcus epidermidis	unknown		

How much gas was produced by *Escherichia coli*?

Did *Staphylococcus epidermidis* ferment? If so, did it produce any gas?

B. Fermentation in yeast

You will be testing just one carbohydrate solution for fermentation in this exercise. Record your data in the table and then complete the table with data obtained by other groups in the class.

Record which substrate solution you are testing:

Hypothesize as to whether you will see much CO_2 production with your assigned solution:

ml of CO_2

minutes	water	glucose	glucose + Na bisulfite	fructose	sucrose	lactose
0	0.0	0.0	0.0	0.0	0.0	0.0
10						
20						
30						
40						
50						
60						

In the space provided, graph the CO_2 production for each solution versus time and draw a curve through each set of points.

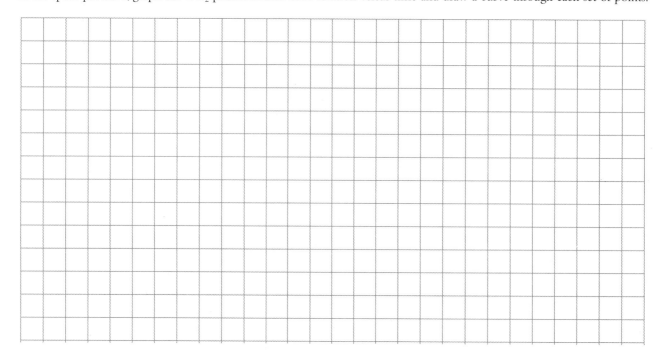

Was your hypothesis confirmed?

Does the rate of fermentation depend on the carbohydrate fermented?

Is there a difference between disaccharides and monosaccharides?

What is the effect of bisulfite on CO_2 production? Why?

C. Succinic dehydrogenase

Sketch or describe the appearance of both the supernatant and the pellet suspension when stained with janus green and viewed under the microscope:

Did you find mitochondria? Which preparation has the most?

Hypothesize which suspension contains the most succinic dehydrogenase:

Complete the table with your observations and analysis:

enzyme source	substrate	color after reaction	interpretation
supernatant	succinate		
supernatant	water		negative substrate control
pellet	succinate		
pellet	water		negative substrate control
supernatant	malonate		
water	succinate		negative enzyme control

In which tubes was the dye reduced? Why wasn't it reduced in the other tubes?

D.   The BZ reaction

Describe how the appearance of the reaction mixture changed after all reactants were mixed together – did a color cycle establish itself? Did you see evidence of the cycles segregating themselves in different micro-environments?

Did gas bubbles form in the reaction? What kind of gas might this be?

Photosynthesis

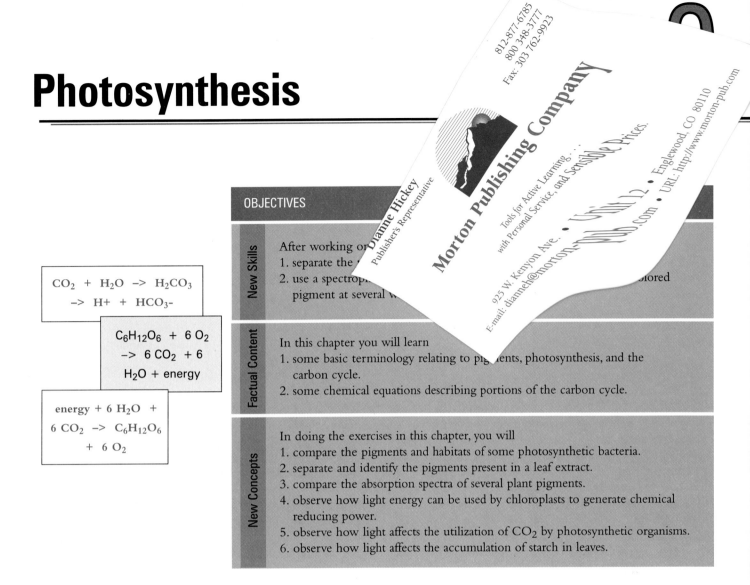

$$CO_2 + H_2O \rightarrow H_2CO_3$$
$$\rightarrow H^+ + HCO_3^-$$

$$C_6H_{12}O_6 + 6 O_2$$
$$\rightarrow 6 CO_2 + 6$$
$$H_2O + energy$$

$$energy + 6 H_2O +$$
$$6 CO_2 \rightarrow C_6H_{12}O_6$$
$$+ 6 O_2$$

OBJECTIVES

New Skills

After working o[...]
1. separate the [...]
2. use a spectrop[...] [c]olored
   pigment at several w[...]

Factual Content

In this chapter you will learn
1. some basic terminology relating to pig[m]ents, photosynthesis, and the carbon cycle.
2. some chemical equations describing portions of the carbon cycle.

New Concepts

In doing the exercises in this chapter, you will
1. compare the pigments and habitats of some photosynthetic bacteria.
2. separate and identify the pigments present in a leaf extract.
3. compare the absorption spectra of several plant pigments.
4. observe how light energy can be used by chloroplasts to generate chemical reducing power.
5. observe how light affects the utilization of CO_2 by photosynthetic organisms.
6. observe how light affects the accumulation of starch in leaves.

INTRODUCTION

The energy source for enzyme-catalyzed reactions is ATP. The energy source to make more ATP is glucose. Where does the energy come from to make glucose?

The chemical reaction that summarizes all of the steps of cellular respiration – glycolysis, the Krebs cycle, and the electron transport chain – simply shows how glucose is oxidized to form carbon dioxide and water, thereby releasing energy to make more ATP:

$$C_6H_{12}O_6 + 6 O_2 \rightarrow 6 CO_2 + 6 H_2O + energy$$

With an outside energy source, this reaction could be reversed to force carbon dioxide and water together to create glucose and oxygen. Many organisms are able to do just that, using energy from the sun to drive the reaction:

$$energy + 6 H_2O + 6 CO_2 \rightarrow C_6H_{12}O_6 + 6 O_2$$

This is the basis of **photosynthesis**. Like cellular respiration, photosynthesis is not a single chemical reaction but a combination of intricate metabolic pathways controlled by many enzymes. The actual enzymatic pathways of photosynthesis are unique to the process – they are not simply respiratory pathways forced to run backwards. Indeed, photosynthetic pathways are always *additions* to respiratory pathways, since *all* cells use glycolysis (and, if aerobic, the Krebs cycle and electron transport chain) to digest glucose. A photosynthetic organism uses the reactions described in this chapter to manufacture carbohydrates, but uses the reactions described in chapter 8 when it later oxidizes those carbohydrates to release their stored energy.

Photosynthesis is the opposite of aerobic respiration in the thermodynamic sense, taking stable, inert molecules and, with the addition of large amounts of solar energy, building them up into monosaccharides. The products of each process are the raw materials for the other. This recycling of the element carbon between carbon dioxide and carbohydrates by living creatures is called the **carbon cycle**. The carbon cycle is a key feature of the ecology and geology of our planet. In the past century, human activity has altered the carbon cycle, because we are now generating more CO_2 by burning fossil fuels than can be absorbed by photosynthesis. As a result, all aspects of the carbon cycle have become very important areas of study. Today's exercises illustrate how light is absorbed and used to generate chemical reducing power, which then stimulates the fixation of CO_2 into organic molecules.

PHOTOSYNTHETIC ORGANISMS

All photosynthetic eukaryotes employ essentially the same kind of chemical process, using water and carbon dioxide as raw materials to produce carbohydrate and oxygen. These eukaryotes differ mainly in what kind of colored **pigment** molecules they use to absorb and trap light. The primary photosynthetic pigments are the three types of **chlorophyll**, designated **chlorophylls *a*, *b*, and *c***. All chlorophylls are various shades of green, which accounts for the most common color in plant and algal tissues. Plants and many algae also contain different **carotenoid** compounds, which act as **accessory pigments**. Accessory pigments do not participate directly in photosynthesis, but they can transfer some of the energy from the light they absorb to chlorophyll. Carotenoids range in color from yellow to orange-brown and are very prominent in brown algae and dinoflagellates, the unicellular algae responsible for poisonous "red tides."

Many kinds of bacteria also contain one form of chlorophyll or another. In the billions of years that they have been on earth, bacteria have developed several mechanisms of photosynthesis, some of which are quite different from that of eukaryotes. Some of the most primitive systems give off sulfur instead of oxygen. Indeed, the photosynthetic **purple** and **green sulfur bacteria** are obligate anaerobes, unable to grow in the presence of oxygen. Those photosynthetic bacteria that do produce oxygen still vary greatly in which pigments they contain, and they can be almost any color: red, brown, green, and blue-green species are all common.

One large and diverse family of photosynthetic prokaryotes is called the **cyanobacteria**. As a group, cyanobacteria are billions of years old, and they were among the first bacteria to give off oxygen as a by-product of photosynthesis. The fossil record suggests that they had a major impact on the chemistry of the entire planet. Because our atmosphere is now about 21% oxygen, there is currently very little pure iron on the surface of the earth — most rusted (oxidized) into various kinds of ores long ago. But locked inside of very old rocks are bits of unoxidized iron, indicating that originally the earth's atmosphere contained very little oxygen. Not until the cyanobacteria appeared did that begin to change. They were so successful and abundant that they completely changed the composition of the air and forced the living world to be dominated by aerobes.

The oxygen-producing reactions that cyanobacteria use in photosynthesis are basically the same as those used by photosynthetic eukaryotes. However, some of the pigments that cyanobacteria use to absorb light are unusual. These pigments, called **phycobilins**, often give the cyanobacteria a blue-green color, which accounts for their common name: **blue-green algae**. However, it must be stressed that not all species of cyanobacteria are blue-green — they can be almost any color, and many species are red. In fact, the only other organisms that contain phycobilin pigments are the eukaryotic red algae; their chloroplasts appear to have evolved directly from endosymbiotic cyanobacteria.

THE CHEMISTRY OF PHOTOSYNTHESIS

The key to photosynthesis is the system of pigments that absorb sunlight so that solar energy can be converted into chemical energy. A pigment is a molecule whose atoms are arranged in such a way that it absorbs some colors of light and reflects other colors. For example, a purple dye might absorb yellow and green light and reflect purple light. When the purple light is reflected toward your eye, you sense the dye as "purple." Chlorophyll reflects green light, so it appears green. However, it is the absorbed light that powers photosynthesis — plants grow better in these colors of light than they do in green light. Accessory pigments are important, because they allow photosynthetic organisms to use some of the energy in colors of light that chlorophyll does not absorb well.

In the chloroplast (or the cell membrane in prokaryotes), the pigment molecules are chemically coupled to enzymes of an electron transport chain that is similar to the one found in mitochondria. However, in chloroplasts the energized electrons used to synthesize ATP come from the light-activated pigments. This makes the system run backwards compared to mitochondria: instead of collecting hydrogen and combining it with oxygen to make water, chloroplasts split water into oxygen and hydrogen. The hydrogen is then stored on molecules of **NADPH** (a phosphorylated derivative of NADH).

When chlorophyll is removed from chloroplasts and their enzyme systems, it cannot participate in photosynthesis, but it still absorbs light energy. Since it cannot use this energy, it re-emits it as heat and lower-energy light. We experience light energy as color: The visible light with the highest energy looks purple to us, and each subsequent color in the **spectrum** (blue, green, yellow, orange, and red, in that order) has lower and lower levels of energy. The re-emission of absorbed light as lower-energy light is called **fluorescence**.

LIGHT AND DARK REACTIONS

The solar energy absorbed during photosynthesis is stored chemically in molecules of ATP and NADPH. Although ATP is the basic source of energy for virtually all of the reactions in a cell, most of the ATP and NADPH generated in photosynthesis is used up in the synthesis of glucose. When it comes to long-term storage, glucose is much more stable than ATP and NADPH. It can also be converted into other kinds of carbohydrates or rearranged into a variety of noncarbohydrate organic compounds.

Because they depend on illuminated pigments for energy, the reactions that generate ATP and NADPH are called **light reactions**. The use of ATP and NADPH to turn CO_2 into glucose does not depend directly on light, so this set of reactions is called the **dark reactions**. However, the dark reactions require a constant supply of ATP and NADPH, and they only occur when these compounds are being actively produced by the light reactions. The two sets of reactions are tightly linked, and in most photosynthetic organisms the dark reactions cease when they run out of ATP and NADPH after a few seconds of darkness.

Just as the light reactions resemble a backwards version of the electron transport chain of cellular respiration, the dark reactions mirror the Krebs cycle. The dark reactions are cyclic, but rather than beginning with a three-carbon compound and using it to generate CO_2 and hydrogen, they take the hydrogens from NADPH and combine them with CO_2 to make a three-carbon compound that can then be used to make glucose. The nature of the dark reactions was elucidated by a man named Calvin, so they are often termed the **Calvin cycle**.

EXERCISES WITH PHOTOSYNTHESIS

Reagents

vegetable extracts	spinach, squash, or corn boiled in 95% ethanol and filtered
chromatography solvent	10% acetone, 5% n-heptane, and 85% mixture of hexane isomers
sucrose solution	0.5 M sucrose
buffer	0.1 M sodium phosphate, pH 6.5
dye solution	0.01% dichlorophenolindolphenol
phenol red solution	0.04% phenol red
IKI solution	

A. Some prokaryotic photosynthetic organisms

Depending on availability, several types of photosynthetic bacteria, such as cyanobacteria and green or purple sulfur bacteria, will be on display.

• Observe the different types of bacterial colonies. Note the colors of each. Can you tell whether they are aerobic or anaerobic? How does this affect their mode of cellular respiration?

B. The chemistry of photosynthesis: Separating photosynthetic pigments

The green color of a plant leaf is actually due to a mixture of pigments: two kinds of chlorophyll (chlorophyll *a* and chlorophyll *b*) and two kinds of carotenoid (**xanthophyll** and **carotene**). All four are nonpolar compounds, but they vary slightly as to exactly how nonpolar they are. Therefore they can be separated from a crude extract by chromatography.

One of the simplest chromatographic techniques is **paper chromatography**. In paper chromatography, a sample mixture is spotted onto one end of a strip of paper. After the solution dries, the tip of the paper is dipped in a solvent. The paper draws up the solvent like a wick, dragging the components of the sample up the paper, too. The parts of the mixture that dissolve in the solvent the best will move the fastest on the paper, while those compounds that dissolve poorly will lag behind. As with any kind of chromatography, the ratio of how far a sample component moves relative to the solvent front (that is, the R_f) is constant for any particular set of conditions.

1. Obtain a strip of chromatography paper. Handle it by the edges, as the oils in your skin can interfere with the chromatography.

2. Mark a line in pencil about 2 cm from one end of the paper. Using a capillary pipet, spot spinach extract two or three times on the pencil line, allowing the extract to dry completely between applications. Your instructor will demonstrate the technique.

3. Place the paper in the chromatography chamber as directed. Be sure that the line of leaf extract is above the level of solvent, or your sample will be washed off the paper entirely. Cover the chamber to contain the solvent fumes. The separation takes 20 to 30 minutes.

4.   Remove your chromatogram. Mark the edge of the solvent front with a pencil and let the paper dry in the fume hood for a few minutes. Then look for the different pigments. Circle these bands with a pencil, as they will fade with time. Measure the distances traveled by each pigment and calculate the R_fs.

5.   The orange and yellow lines at the top are the carotenoids carotene and xanthophyll, respectively. Below these pigments is a large blue-green spot followed by a small, yellow-green band. These are chlorophyll *a* and chlorophyll *b*. Compare the colors of these pigments to the other extracts (the squash extract is rich in carotene and the corn extract is rich in xanthophyll). Which pigment seems to be the major component of this mixture? Is this what you would expect based on the chemistry of the light reactions?

C.   The chemistry of photosynthesis:
Absorption spectra of photosynthetic pigments

Previously you have used the spectrophotometer to measure the concentration of a substance in a solution. You can also use a spectrophotometer to measure a substance at several wavelengths to find out how well it absorbs different colors of the spectrum.

1.   Set up the spectrophotometer as described in chapter 5. Set the wavelength at 400 nm for the first measurement.

2.   Fill a spectrophotometer tube with diluted spinach extract and a second with ethanol. The ethanol will be the blank (all of the test solutions are in ethanol). Even though you know from exercise B that the spinach extract is a mixture, consider it to be pure chlorophyll *a* for this experiment – that is the most abundant pigment in the mixture.

3.   Zero the spectrophotometer at 400 nm using the blank and measure the absorbance of the chlorophyll.

4.   Adjust the wavelength to 420 nm and re-zero the instrument. Read the chlorophyll absorbance again. Continue taking readings every 20 nm until you reach 700 nm, then stop. You must re-zero every time you change the wavelength.

5.   Repeat the process for squash extract (carotene) and corn extract (xanthophyll). Tabulate all of the results.

6.   On a single graph, plot all three absorption spectra. Compare the positions of the peaks for each pigment. From this graph, which wavelengths would you expect to promote photosynthesis and plant growth the best? Which wavelengths would be the worst? What colors do these wavelengths represent?

D.   Light reactions

In one of the light reactions water is split into hydrogen and oxygen. The hydrogen is stored in NADPH, which is generally used to reduce CO_2 to make organic molecules. (The light-induced splitting of water is sometimes called the **Hill reaction** after its discoverer, R. L. Hill.) As you recall from last week, we can intercept the hydrogens stored on NADH during cellular respiration and use them to decolor the reducible dye dichlorophenolindolphenol. We can do the same thing with chloroplasts – use them to reduce dye molecules. In this case we won't need succinate as an energy source; we will use light.

1.   Your lab instructor will begin by homogenizing about 40 g of spinach leaves in a blender with 200 ml of cold sucrose solution. The chloroplasts will then be pelleted by centrifugation and the supernatant discarded. The chloroplasts will be resuspended in 75 ml cold sucrose solution and should be kept on ice until you are ready to work with them.

2.   Prepare two test tubes as follows:

   2 ml buffer
   2 ml dye solution
   2 ml chloroplast suspension
   3 ml water

Wrap one tube in foil to exclude light.

3.   Prepare two control tubes with:

   2 ml buffer
   5 ml water

To the first add 2 ml of dye solution and to the second add 2 ml of chloroplast suspension.

4.   Place all tubes in a large beaker of cold water. Put them in front of a strong light until the uncovered tube from step 2 (with both chloroplasts and dye) is the same color as the tube from step 3 containing chloroplasts but no dye (about 20 minutes).

5.   Uncover the foil-wrapped tube and compare the colors in all the tubes. Tabulate your results. In which tube(s) did reduction occur? Why?

E.   Dark reactions: The uptake of CO_2

We were able to monitor fermentation by measuring the CO_2 that was given off by the process. Since the dark reactions use

CO_2, we ought to be able to observe them absorbing the gas. This is far more complex than the system we used for fermentation, however. For one thing, photosynthesis releases oxygen as it uses carbon dioxide, so this would offset any measurable decrease in gas volume. (Why wasn't there an analogous problem during fermentation?) For water plants, algae, or bacteria, the problem is further complicated by the fact that the CO_2 must first dissolve in the watery medium before it can be used for photosynthesis. Actually, carbon dioxide is too nonpolar to dissolve very well in water, but it will react with water to form **carbonic acid**:

$$CO_2 + H_2O \rightarrow H_2CO_3 \rightarrow H^+ + HCO_3^-$$

This means that solutions with a high concentration of carbon dioxide have a lower pH than solutions with a low concentration of carbon dioxide. (Several gases, including NO_2 and SO_2, form acids on reacting with water. This is the principle behind the formation of acid rain, and it is one of the reasons that burning excess fossil fuels is an environmental problem.) Because of this, we can observe the absorption of CO_2 from a solution by using a colored pH indicator. The method is not quantitative, as was our device for measuring fermentation, but it does illustrate the basic principle.

1. Dilute about 20 ml of 0.04% phenol red solution four-fold in a flask and blow into the solution with a straw until the solution turns yellow. Excess carbon dioxide is not harmful, but it will lengthen the time of the exercise.

2. Place some of the yellow solution into two depressions of a spot plate.

3. Place some algae in one of the depressions. Shine a light on the spot plate.

4. Record the time needed for the solution with the algae to turn pink. If the control (the solution without algae) turns pink, start over. Where in the spot plate does the solution first begin to turn pink – near the algae or away from it?

F. Dark reactions: Light dependence of carbohydrate accumulation

Plants convert the organic products of photosynthesis into glucose. As the plant accumulates glucose, it polymerizes the monosaccharide into starch. That is, prolonged photosynthesis leads ultimately to a buildup of starch. Leaves can be tested for this with IKI.

Period One

1. Your instructor will have some plants available. Follow the instructor's directions in applying a mask of paper or foil to one of the leaves. You may make the mask any shape you wish.

2. Place the plant under well-lit conditions until the next lab period.

Period Two

1. Remove your leaf from the plant and remove the mask. Sketch the leaf, showing what part was masked and what part was exposed to light.

2. Place the leaf in boiling ethanol (in a fume hood!) to extract the pigments and bleach the leaf.

3. Place the bleached leaf in a petri dish and cover with IKI solution. Where is the starch in the leaf? How does this relate to the portion covered by the mask?

IMPORTANT TERMS

accessory pigment	carotenoid	fluorescence	photosynthesis
blue-green algae	chlorophyll	green sulfur bacteria	phycobilin
Calvin cycle	chlorophyll *a*	Hill reaction	pigment
carbon cycle	chlorophyll *b*	light reactions	purple sulfur bacteria
carbonic acid	cyanobacteria	NADPH	spectrum
carotene	dark reactions	paper chromatography	xanthophyll

Laboratory Report Photosynthesis

FACTUAL CONTENT

Define the following terms. Brief definitions are given in the chapter, but you may also wish to consult your lecture notes and text for a better idea of how the terms are used. Pay close attention to those words that look similar but have different meanings — you will want to be able to tell them apart if they appear next to each other on a multiple-choice exam.

accessory pigment	fluorescence
blue-green algae	green sulfur bacteria
Calvin cycle	Hill reaction
carbon cycle	light reactions
carbonic acid	NADPH
carotene	paper chromatography
carotenoid	photosynthesis
chlorophyll	phycobilin
chlorophyll *a*	pigment
chlorophyll *b*	purple sulfur bacteria
cyanobacteria	spectrum
dark reactions	xanthophyll

Write the chemical equation that summarizes all of the steps of cellular respiration — glycolysis, the Krebs cycle, and the electron transport chain:

Write the chemical equation that summarizes photosynthesis:

Write the chemical equation that represents the reaction of carbon dioxide with water to form carbonic acid:

A. Some prokaryotic photosynthetic organisms

How many different kinds of bacteria are on display? What colors are they?

Can you tell whether they are aerobic or anaerobic? How does this affect their mode of cellular respiration?

B. The chemistry of photosynthesis: Separating photosynthetic pigments

Record the distance traveled by the solvent front on your chromatogram:

Complete the table for the colored bands you obtained on your chromatogram:

color	distance	R_f	pigment identity

Which pigment seems to be the major component of this mixture? Is this what you would expect based on the chemistry of the light reactions?

C. The chemistry of photosynthesis: Absorption spectra of photosynthetic pigments

Complete the table for the measured absorptions of each pigment:

wavelength	chlorophyll	carotene	xanthophyll
400 nm			
420 nm			
440 nm			
460 nm			
480 nm			
500 nm			
520 nm			
540 nm			
560 nm			
580 nm			
600 nm			
620 nm			
640 nm			
660 nm			
680 nm			
700 nm			

In the space provided, graph the absorption spectrum for each pigment and draw a curve through each set of points.

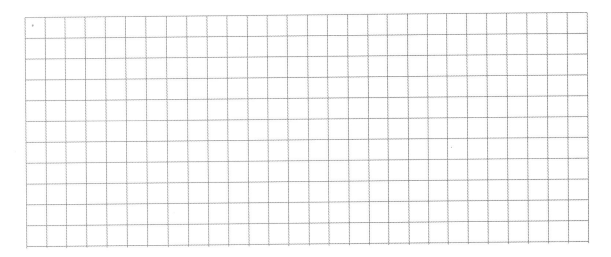

Which wavelengths would you expect to promote photosynthesis and plant growth the best? Which wavelengths would be the worst? What colors do these wavelengths represent?

D. Light reactions

Complete the table with your observations and analysis:

chloroplasts	dye	light	color after reaction	interpretation
present	present	yes		
present	present	no		
absent	present	yes		negative chloroplast control
present	absent	yes		negative dye control

In which tubes was the dye reduced? Why wasn't it reduced in the other tubes?

E. Dark reactions: The uptake of CO_2

Photosynthesis releases oxygen as it uses carbon dioxide, offsetting any measurable decrease in gas volume. Why wasn't there an analogous problem measuring fermentation in chapter 8?

Record the time needed for the solution with the algae to turn pink:

Where in the spot plate does the solution first begin to turn pink – near the algae or away from it?

F. Dark reactions: Light dependence of carbohydrate accumulation

Sketch the leaf, showing what part was masked and what part was exposed to light:

Where is the starch in the leaf? How does this relate to the portion covered by the mask?

Cell Division

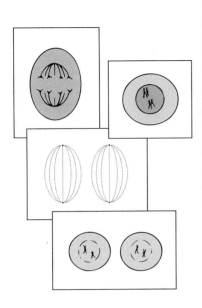

INTRODUCTION

Cells reproduce by dividing in two. In eukaryotic cells this process is complex, because all of the organelles must be duplicated and divided up as well. The most important organelle is the nucleus, since it contains the genetic material for the cell. The process of nuclear division is called **mitosis**. During mitosis the genetic material, packaged as distinct chromosomes, is distributed among the new nuclei.

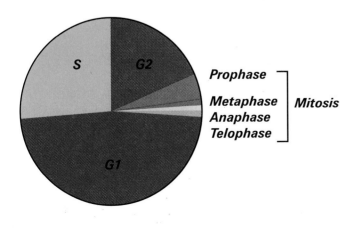

Mitosis is one phase of the life cycle of the cell. The entire **cell cycle** is divided into four main parts: The first phase is a **gap** between reproductive activities and is designated G_1. G_1 phase cells are doing anything but reproducing – this is the "normal" part of the cell's life and is usually the longest part of the cell cycle (see figure 10.1). After G_1 comes the **S**, or **synthesis**, phase. At this time normal activity stops and the nuclear DNA replicates itself. Note that replication is not part of mitosis. After the S phase comes another gap (G_2) before mitosis. During G_2, proteins necessary for mitosis are assembled. Some of these proteins modify the cell's internal framework, or **cytoskeleton**, into a special **spindle** apparatus. The **spindle fibers** do the work of separating and moving the chromosomes during mitosis. At the end of mitosis, **cytokinesis**, or cytoplasmic division, occurs, giving rise to two separate cells. After cytokinesis, both new cells start out in the G_1 phase again.

Figure 10.1 *The phases of the cell cycle*

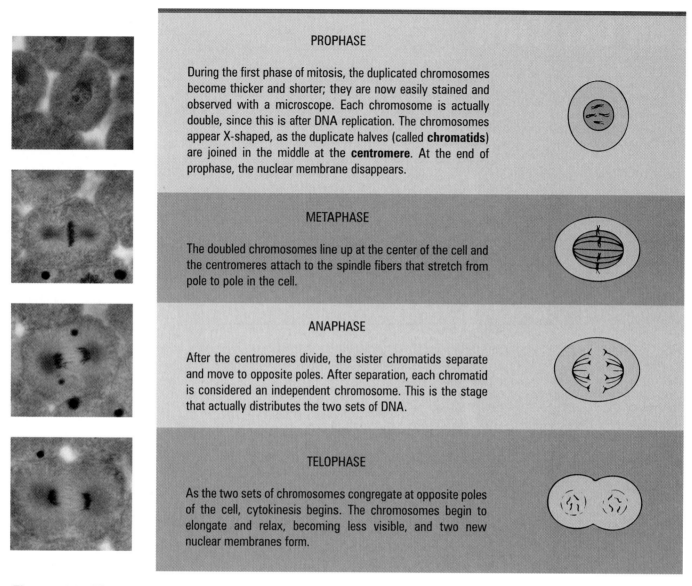

PROPHASE

During the first phase of mitosis, the duplicated chromosomes become thicker and shorter; they are now easily stained and observed with a microscope. Each chromosome is actually double, since this is after DNA replication. The chromosomes appear X-shaped, as the duplicate halves (called **chromatids**) are joined in the middle at the **centromere**. At the end of prophase, the nuclear membrane disappears.

METAPHASE

The doubled chromosomes line up at the center of the cell and the centromeres attach to the spindle fibers that stretch from pole to pole in the cell.

ANAPHASE

After the centromeres divide, the sister chromatids separate and move to opposite poles. After separation, each chromatid is considered an independent chromosome. This is the stage that actually distributes the two sets of DNA.

TELOPHASE

As the two sets of chromosomes congregate at opposite poles of the cell, cytokinesis begins. The chromosomes begin to elongate and relax, becoming less visible, and two new nuclear membranes form.

Figure 10.2 *The phases of mitosis*

The whole point to mitosis is to provide each of the daughter cells with a complete set of chromosomes identical to the chromosomes of the original mother cell. It is a method of distributing the DNA that was duplicated during the S phase of the cell cycle.

MITOSIS

Mitosis is traditionally subdivided into four phases. These are described in figure 10.2.

MEIOSIS

Mitosis generates two cells that are genetically identical to each other and to the parent cell. This is not always desirable. When a sperm and an egg unite in sexual reproduction, each must contain only half the normal amount of chromosomes or the fertilized egg will have too many chromosomes. Eggs and sperm are formed in a special type of mitosis called **meiosis**. (Meiosis can also lead to haploid spores and individuals in some species, as we will see in chapter 18.)

Eukaryotic chromosomes work on a kind of "buddy system" where each chromosome has a partner. The chromosomes of each pair have the same types of genes, but they are not necessarily identical. For example, one pair may have the genes for eye color. Both chromosomes would have an eye color gene in the same place, but one might have a gene for blue eyes while the other has one for brown. These pairs are called **homologous chromosomes**. We will explore the genetic implications of the

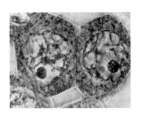

PROPHASE

This is similar to prophase in mitosis, except that the double chromosomes line up (**synapse**) with their homologs. At this time **crossing over** may occur, in which segments of DNA are exchanged between the different chromatids in each pair.

METAPHASE I

The synapsed homologs line up at the center of the cell. Unlike mitotic metaphase, here there is a *pair* of doubled chromosomes on each spindle fiber.

ANAPHASE I

The homologous pairs separate. At this point the centromeres do not divide, and the *sister chromatids remain together*.

TELOPHASE I

The separated chromosomes form two clusters at opposite poles of the cell. Usually cytokinesis takes place, forming two cells. Note that the chromosomes in each cell are still doubled: two chromatids joined by a centromere. There are half as many of these in each cell as there were in prophase.

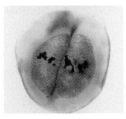

METAPHASE II

The doubled chromosomes line up at the center of each cell.

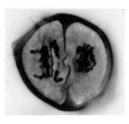

ANAPHASE II

This time the centromeres do divide, and the sister chromatids separate.

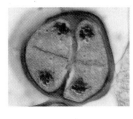

TELOPHASE II

There are now four cells with single chromosomes. Each has half the normal number of chromosomes. Since the homologous pairs were not originally identical, the daughter cells from meiosis are not all genetically identical.

Figure 10.3 *The phases of meiosis*

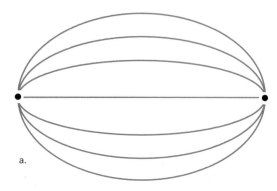

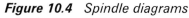

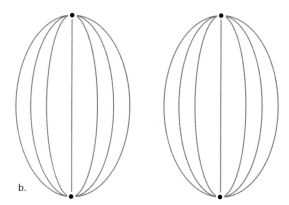

Figure 10.4 *Spindle diagrams*
a. mitotic spindle
b. double spindle for metaphase II and anaphase II

fact that chromosomes come in pairs in the next chapter. For now we will focus on how homologous chromosomes are separated in meiosis.

Homologous chromosomes are not the same as sister chromatids – sister chromatids are identical chromosomes formed during the S phase of the cell cycle. After S phase there are pairs of doubled chromosomes. In mitosis, the fact that the chromosomes come in pairs is ignored: the doubled chromosomes are just split into single chromatids and divided equally among the two new cells. However, in meiosis the pairs are first divided up and then the sister chromatids separate, creating a total of four daughter cells. Since there are two rounds of separation of chromosomal material in meiosis, there are more stages (figure 10.3).

EXERCISES WITH MITOSIS AND MEIOSIS

A. Duration of mitotic phases

1.   Examine a prepared slide of an onion root tip. In roots, cell division occurs in a zone just behind the very tip. Under high power you should be able to see cells in all phases of mitosis. Identify as many phases as you can.

2.   Adjust the slide so that the field covers an area of intense mitotic activity. Count the total number of cells in the field and the number in each phase of mitosis. Tabulate your results with the rest of the class. Assuming a cell cycle of 24 hours, the length of each phase can be determined by a ratio:

$$\frac{\text{duration of phase}}{24 \text{ hours}} = \frac{\text{number of cells in that phase}}{\text{total number of cells}}$$

How long is each phase? (Published values are: prophase, 71 minutes; metaphase, 6.5 minutes; anaphase, 2.4 minutes; telophase, 3.8 minutes; and interphase, the remaining 22 hours 36 minutes. Class-determined values will probably not agree with these exactly, but they should be similar.)

B. Simulating mitosis

The events of mitosis and meiosis can be very confusing. You and your partner will be given a kit so that you can simulate chromosome movement during the two types of nuclear division. Work through the simulation until you feel you understand the differences between mitosis and meiosis.

Your model cell will contain two pairs of chromosomes. In the first pair, one chromosome will be made of red beads and the other of yellow beads. The second pair will consist of a blue and a green chromosome. Each chromosome is made by sticking a string of beads in each end of a magnetic "centromere;" your instructor will demonstrate. The centromere need not be in the exact center of the chromosome, but both chromosomes of each pair must be the same size and shape.

1.   Draw a circle on your bench top with chalk to represent a G$_1$ nucleus. Place one chromosome of each color in the nucleus and keep this model as a reference.

2.   Draw a "prophase nucleus" with chalk. Place four doubled chromosomes in this circle. The chromatids of each doubled chromosome should be identical in shape, size, and color. The magnetic centromeres will hold the chromatids together in the proper **X** shape.

3. Next, draw a spindle apparatus (figure 10.4a). For metaphase, the doubled chromosomes are lined up on four separate spindle fibers at the center of the spindle.

4. To simulate anaphase, pull the magnets apart and separate the chromatids of each chromosome.

5. When you are done, you should have two groups of single chromosomes, representing the nuclei of the new daughter cells. How does each new nucleus compare to the reference nucleus? How do they compare to each other? Draw the G_1 reference nucleus in your notes along with the stages of mitosis so that you have a record of how this result was obtained.

C. Simulating meiosis

1. Begin again as you did for the mitosis simulation by making a G_1 nucleus as a reference.

2. For prophase, line up each chromosome with its homolog in the nucleus (i.e., red with yellow and blue with green). The two doubled chromosomes of each pair should lie next to each other so that in the center, two homologous chromatids of contrasting color are touching — red next to yellow and blue next to green. This is where crossing-over occurs. Exchange the beads on the tips of the touching chromatids at one end (e.g., one red chromatid should have a few yellow beads at one end and the adjacent yellow chromatid should have a red tip). Since crossing-over just affects one chromatid in each chromosome, each homologous pair should now contain four different chromatids (e.g., all red, red with a little yellow, yellow with a little red, all yellow).

3. In metaphase I the synapsed pairs remain together, so the spindle should have a pair of doubled chromosomes on each of two fibers.

4. In anaphase I the members of each homologous pair separate to the poles.

5. Next draw a double spindle below the first (figure 10.4b). Line the chromosomes up at the center; there should be no homologs to pair up.

6. Separate the chromatids (that is, pull apart the magnets) for anaphase II.

7. You should now have four sets of single chromosomes. How does each new nucleus compare to the reference nucleus? How do they compare to each other? Draw the G_1 reference nucleus in your notes along with the stages of meiosis so that you have a record of how this result was obtained.

IMPORTANT TERMS

anaphase	cytokinesis	meiosis	spindle fibers
cell cycle	cytoskeleton	metaphase	synapse
centromere	gap 1 (G_1)	mitosis	synthesis (S)
chromatid	gap 2 (G_2)	prophase	telophase
crossing over	homologous chromosomes	spindle	

Laboratory Report Cell Division

FACTUAL CONTENT

Define the following terms. Brief definitions are given in the chapter, but you may also wish to consult your lecture notes and text for a better idea of how the terms are used. Pay close attention to those words that look similar but have different meanings — you will want to be able to tell them apart if they appear next to each other on a multiple-choice exam.

anaphase

meiosis

cell cycle

metaphase

centromere

mitosis

chromatid

prophase

crossing over

spindle

cytokinesis

spindle fibers

cytoskeleton

synapse

gap 1 (G_1)

synthesis (S)

gap 2 (G_2)

telophase

homologous chromosomes

A. Duration of mitotic phases

Record the number of cells you counted in each phase of mitosis in the table:

phase	number of cells
interphase	
prophase	
metaphase	
anaphase	
telophase	
total	

Complete the table using the combined data for the whole class and calculate the length of each phase based on the entire sample:

phase	number of cells	duration
interphase		
prophase		
metaphase		
anaphase		
telophase		
total		24 hours

B. Simulating mitosis

Draw the G_1 reference nucleus:

Draw and label the stages of mitosis:

How does each new nucleus compare to the reference nucleus? How do they compare to each other?

C. Simulating meiosis

Draw the G_1 reference nucleus:

Draw and label the stages of meiosis:

How does each new nucleus compare to the reference nucleus? How do they compare to each other?

Genetics

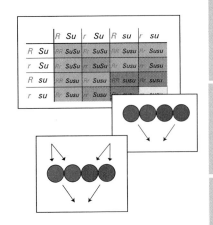

INTRODUCTION

A **gene** is an individual unit of heredity. Physically, a gene is any particular stretch of DNA on a chromosome that codes for a specific trait. Since traits can have different possible forms (e.g., hair can be dark or blond, eyes can be blue or brown), it is clear that there can be several different forms of a gene for a single trait. Alternative forms of a gene are called **alleles**.

One hundred years ago, Gregor Mendel discovered that genes come in pairs and that some alleles are able to mask others. These discoveries explained a great deal and laid the groundwork for the modern science of genetics. Genetics has grown tremendously since Mendel, but the principles he elucidated are still important. This chapter focuses on Mendel's basic principles.

SIMPLE DOMINANCE

Many genetic traits have been studied in corn in order to facilitate the breeding of improved crops. It is known, for example, that when dark corn is crossed with yellow corn, the offspring are usually dark – the yellow allele seems to just disappear. Because of this, the dark allele is said to be **dominant** over the yellow allele, and the yellow allele is said to be **recessive**.

The offspring from two different purebred strains is called a **hybrid**. Dark corn formed by crossing dark corn with yellow corn is an example of a hybrid. When hybrid dark corn is crossed with hybrid dark corn, a surprise occurs: the next generation is not all dark – 25% of the kernels are yellow. This means that the dominant allele does not destroy the recessive allele, it merely masks it for a generation.

Mendel found that hybrid crosses such as these always give rise to offspring of which 75% show the dominant trait and 25% show the recessive. He was able to figure out that this meant that each individual carries two genes for each trait. Here's how this works for the corn:

The dark corn has two genes for color. Since it is purebred, the alleles are the same (that is essentially what purebred means). The dark being dominant, we abbreviate this allele with a capital letter, in this case, *R*. Thus the two genes in the pure dark corn are *RR*. *RR* is called the **genotype** of this corn. The pure yellow corn has two recessive alleles for a genotype of *rr*. When both genes in an individual are the same allele, the individual is called **homozygous**. The term *homozygous* refers directly to the genotype of the individual and is therefore more precise than the term we used earlier, *purebred*, which technically means that the individual is bred from a long line of individuals that all exhibited one form of a particular trait. Since we will be discussing genotypes from now on, we will use the term *homozygous* instead of *purebred*.

When the two strains of corn are crossed, each parent contributes a single gene to the offspring. The paired genes for each trait reside on homologous chromosomes; as we saw in the last chapter, it is meiosis that allows these pairs to be separated prior to sexual reproduction.

Homozygous parents have the same type of allele on both homologs, and that is the only allele that they can pass on to their children. In this case, the dark corn passes on the *R* allele and the yellow corn passes on the *r* allele. The genotype of all of the children is *Rr*. When the genes in an individual are of different alleles, the individual is called **heterozygous**. In cases of simple dominance, heterozygous individuals show the dominant trait. The heterozygous corn kernels are all dark. This means that two different genotypes look alike: *RR* and *Rr* are both dark. The appearance of an individual is called the **phenotype**. The phenotype is determined largely by the genotype, but several different genotypes may give the same phenotype.

When the heterozygous corn is crossed with heterozygous corn, four types of offspring are equally likely:

RR	(*R* from each parent)
Rr	(*R* from mother, *r* from father)
Rr	(*R* from father, *r* from mother)
rr	(*r* from each parent)

Thus we would expect 25% of the children to be *RR*, 50% to be *Rr*, and 25% to be *rr*. Both the *RR* and the *Rr* genotypes give the dark phenotype, so the ratio of children is 75% dark to 25% yellow. Mendel summarized these patterns in his **law of segregation**: the paired alleles of each parent will separate (segregate) when the parents form gametes and will create new combinations when the gametes randomly fuse to form zygotes.

DIHYBRID CROSSES AND PUNNETT SQUARES

Mendel's theory works for traits involving a single gene in which one allele is dominant over another. By studying more than one trait at a time, Mendel was able to extend his analysis to more complicated situations. A cross between individuals that are heterozygous for a single trait is called a **monohybrid cross**. If they are heterozygous for two traits, the cross is a **dihybrid cross**.

For an example of a dihybrid cross in corn, consider a second trait designated *Su*. (Since there are too many genetic traits to be designated by the 26 letters of the alphabet, many traits are given two-letter abbreviations. Don't confuse the two letters of a single trait with the two alleles of a genotype!) The *Su* trait determines whether the corn kernel stores carbohydrate as sugar or starch and is easily recognized, since the sugary kernels look wrinkled while the starchy kernels are smooth. Smooth and starchy (*Su*) is dominant over wrinkled and sweet (*su*), so the possible genotype/phenotype combinations are:

SuSu	(starchy and smooth homozygote)
Susu	(starchy and smooth heterozygote)
susu	(sweet and wrinkled homozygote)

A corn kernel heterozygous for both color and carbohydrate would have the genotype:

RrSusu

and would look dark and smooth. The color and carbohydrate genes are independent of each other and can be inherited in any combination, so a dihybrid cross has many more possibilities. The possible combinations of genes each parent could contribute to the next generation are shown in figure 11.1.

Any set from one parent could match up with any set from the other. The easiest way to sort this out is by drawing a **Punnett square**, a table showing the possibilities from one parent across the top and the other parent down the side (figure 11.2).

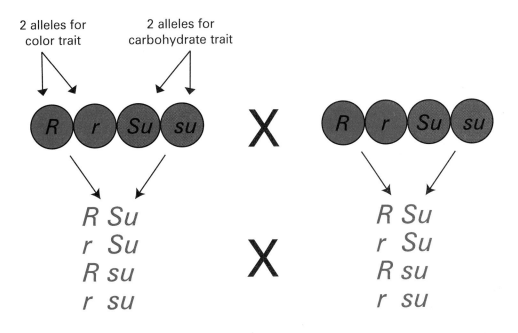

Figure 11.1 *Possible allele combinations from heterozygous dihybrid parents*

	R Su	r Su	R su	r su
R Su	RR SuSu	Rr SuSu	RR Susu	Rr Susu
r Su	Rr SuSu	rr SuSu	Rr Susu	rr Susu
R su	RR Susu	Rr Susu	RR susu	Rr susu
r su	Rr Susu	rr Susu	Rr susu	rr susu

Figure 11.2 *Punnett square for a dihybrid cross*

There are sixteen combinations. Any with at least one dominant allele for a trait will show the dominant phenotype for that trait, so we can narrow the sixteen genotypes down to four phenotypes:

9   *R–Su–*   (dark and smooth)
3   *R–susu*   (dark and wrinkled)
3   *rrSu–*   (yellow and smooth)
1   *rrsusu*   (yellow and wrinkled)

Whenever two parents who are both heterozygous for two traits are crossed, the offspring will show this 9:3:3:1 dihybrid ratio. If they are not both heterozygous for both traits, other ratios are possible.

Notice that the dihybrid cross gives us new combinations of phenotypes – in this case, dark / wrinkled and yellow / smooth – that were not present in the original parents. This is because the two traits are *independent* of each other. Mendel formulated the **law of independent assortment** to describe how these new combinations arise. We saw in the last chapter how independent assortment occurs during meiosis.

THE HARDY-WEINBERG EQUATION

The fact that one allele is dominant over another allele does not mean that that allele is more common than the recessive allele or even that it will become more common over time – *dominance refers only to which phenotype will be expressed in a heterozygote.* Dominance and recessiveness have no direct bearing on allele frequencies.

The only way to study the distribution of traits in an entire population is by statistics. At the beginning of the twentieth century two geneticists, working independently, used statistical principles to develop theoretical calculations of gene frequencies. They pointed out that in an ideal population, allele frequencies would settle into an equilibrium in the absence of outside forces. An "ideal population" is defined as a large population whose members mate randomly (i.e., all possible genetic crosses occur just as random statistics would predict). By "outside forces" we mean some factor that would force the population to change genetically: natural selection, mutation, or an influx of new genes by immigration. This theory of genetic equilibrium, including the equations that describe it, is called the **Hardy-Weinberg theorem** after the two scientists.

Hardy and Weinberg derived an algebraic equation to relate the frequency of one allele to another. If a trait has only two alleles, a dominant and recessive, then some fraction of all of the alleles for that trait would be dominant and the rest would be recessive. Obviously, the two fractions must add up to 100%. So, if we designate the fraction of alleles that are dominant as *p* (it could be any number) and the fraction that are recessive as *q*, then together they must add up to a whole:

$$p + q = 1$$

That describes the total frequency of each allele. But every individual has two alleles in their genotype. To reflect this we have to consider all of the possible combinations of the alleles taken two at a time. Since we have already said that mating is random, any allele is free to pair up with any other allele, and the odds of doing so are based solely on the frequencies of the alleles. The frequency of a combination is found by multiplying the frequencies of the parts. (For example, the chance of getting a "heads" on a coin toss is 50% or 0.5. The chance of getting two heads if you toss two coins together is 0.5 times 0.5 = 0.25, 25%.) Therefore we need to square both sides of the equation:

$$(p + q)(p + q) = 1$$
or
$$p^2 + 2pq + q^2 = 1$$

where:

p^2 = the fraction of dominant homozygotes
$2pq$ = the fraction of heterozygotes, and
q^2 = the fraction of recessive homozygotes

Of course, the dominant homozygotes and the heterozygotes have the same phenotype and are indistinguishable, but the recessive homozygotes can be easily picked out. In our corn example, both the dominant homozygotes and the heterozygotes would be dark colored, but the recessive homozygotes would be yellow. The equations allow you to use the recessive allele frequency to figure out the rest. Let's say that 36% of a population shows the recessive phenotype. That means that $q^2 = 0.36$. Therefore $q = 0.6$. Since $p + q = 1$, p must equal 0.4. Therefore 16% ($0.16 = 0.4^2$) are dominant homozygotes and 48% [$2pq = 2(0.4)(0.6) = 0.48$] are heterozygotes.

Strictly speaking, the Hardy-Weinberg equation works only in ideal cases. When applied to a real population, the numbers will be at best a rough estimate of the actual allele frequencies. The accuracy of the estimates will depend on how close the population is to being "ideal" (large and with random mating) and how strong the "outside forces" are that affect the population.

CODOMINANCE AND MULTIPLE ALLELES

Not all common traits are a matter of simple dominance. **Blood types**, for example, exhibit a property called **codominance**. As you probably know, there are four major blood groups – A, B, AB, and O. These four types are determined by the **multiple alleles** of a single gene. Blood type also is influenced by a second gene for the **Rh factor**. The Rh factor is a simple Mendelian trait in which the positive blood type is dominant over the negative. Rh factor is inherited independently from the A,B,O type.

Blood type describes a particular protein on the surface of red blood cells. The allele I^A produces a type-A protein and the allele I^B produces a type-B protein. Both of these alleles are dominant over i, which produces no protein at all. The presence of two dominant alleles is what makes blood type an example of codominance. Although both dominant alleles overpower the recessive allele, neither is able to overpower the other dominant allele. In a heterozygote with both dominant alleles, both of the alleles are fully expressed. Thus the six possible genotypes for blood type give four phenotypes:

$I^A I^A$ or $I^A i$	Type A
$I^B I^B$ or $I^B i$	Type B
$I^A I^B$	Type AB
ii	Type O

Type A blood has A proteins (often called A **antigens** because they can generate an antibody response) on its red blood cells. Type B has B antigens, type AB has both A and B antigens, and type O has no antigens. This is important for blood transfusions, because the body can recognize antigens that are foreign. If type A blood is given to a type B person, the A antigens will be detected by the recipient. The recipient's immune system will use the antigen as a target and try to destroy all of the A blood cells. Similarly, type B blood cannot be given to a type A recipient. Type O blood is valued by blood banks because it has no antigens that can be recognized by anyone. It is therefore called the **universal donor**. Which type would be the **universal recipient**?

Blood types are sometimes used to help settle paternity cases because they are fairly easy to determine and cannot be faked. For example, it is impossible for a man with type AB blood to be the father of a child with type O blood. Why? What other combinations are impossible? What combinations are ambiguous?

CHI-SQUARE ANALYSIS

Theoretical genetics makes predictions based on probabilities, but in the real world these predictions do not always come true. For example, we know that the probability of a human child being a girl is 50%, but it is not unusual for a family to have three or four daughters and no sons. The 50% prediction is based on statistics, and it is not very reliable in specific cases or small populations. Although geneticists try their best to use large samples in their experiments, the perfect Mendelian ratios are never seen exactly.

It is possible for statisticians to predict how likely their predictions are to come true. The **chi-square test** was devised to tell if experimental variance with a hypothesis is due simply to chance or if it is due to a real problem with the hypothesis. It essentially predicts how often one should expect a particular level of deviation in one's data.

To perform a chi-square test, you compare your observed results with the results of your hypothetical prediction and total the numerical differences in a special statistical index, the chi-square value. Statisticians have already compiled tables showing how likely a given amount of difference is to occur. By convention, any observed difference that is at least 5% likely to occur by chance is considered normal and does not automatically indicate a problem with the hypothesis. That does not mean that the hypothesis is proven, it just means that the statistics are not strong enough to disprove it.

Chi-square analyses have become a basic tool for geneticists. Because the theories of Mendel are now well-established by decades of observations and experiments, chi-square results are not interpreted as reflecting on the accuracy of Mendel's work. Instead, if a particular experiment shows a significant deviation from a Mendelian hypothesis, the geneticist looks for some reason why the cross being studied might not conform to Mendel's rules: Have all the alleles been identified? Is there codominance? Might the genes under study interact in unusual ways? Such reasoning led to the discovery of **linkage groups** at the beginning of the twentieth century. A linkage group is a set of genes that are inherited together and do not sort independently. A dihybrid cross involving linked genes does not give a 9:3:3:1 ratio because the alleles of the two different genes cannot make all of the possible combinations shown in figure 11.1. Instead, the ratio comes closer to 1:0:0:1. The implication that a linkage group was a package of genes that were physically attached to each other was one of the observations that led to the realization that genes were located on chromosomes. In such situations, new combinations of the linked genes can only arise when crossing over occurs.

115

EXERCISES WITH GENETICS

A. Simple dominance, mono- and dihybrid crosses, and Punnett squares

Punnett squares are an easy way to keep track of what is happening in a single genetics cross. Sometimes it is necessary to trace genotypes over several generations in a **pedigree**, or family tree. Use both of these tools to work out the following hypothetical problems:

1. Draw Punnett squares for the following crosses:

   RR x *rr*
   Rr x *Rr*
   RR x *Rr*
   Rr x *rr*

2. A woman has type A blood but her father is type O. What is her genotype? If she marries a type O man, what proportion of their children would you expect to have type O blood? What proportion should have type A? Type B?

3. Mary has brown eyes (*BB*). Her father was blue-eyed and her mother was brown-eyed. She marries a blue-eyed man whose parents were both brown-eyed. Mary has a son with blue eyes. Can you figure out the genotypes of everyone in the family? Which ones can you be sure of? Draw a pedigree showing their genotypes. Label the genotypes you are unsure of as "*B?*".

4. *Megacoccus multichromatium* is a strange creature known for its colorful shells and also for its ability to store enormous amounts of sucrose in its body. A scientist obtained two purebred strains of *M. multichromatium*, the first of which was dark brown and always gave dark brown progeny, and the second of which was yellow and always gave yellow progeny. He mated these and obtained offspring that were all dark brown. However, when he crossed these second generation (heterozygous) brown creatures, the 32 offspring were colored as follows:

   18 dark brown
   6 red
   6 light green
   2 yellow

How many genes (not alleles!) control shell color in *M. multichromatium*? Draw Punnett squares for both the original cross between brown and yellow and the second cross between heterozygotes.

B. Simple dominance in human traits

Not all traits are cases of simple dominance. Many traits are the result of several genes working together, and some traits have more than two possible alleles. In these complicated systems, there may be many more than just two possible phenotypes. Often there is an infinite variety of continuously variable possibilities. This is seen in traits such as human height or skin color (of course, both of these traits are also influenced by the environment). However, several simple traits have been identified in humans that illustrate Mendelian genetics quite well. A few are listed here.

Acondroplastic dwarfism The dominant allele causes dwarfed arms and legs; normal proportions are recessive.

Albinism Normal pigment production is dominant (*A*) over albinism (pale skin and hair, pink irises).

Attached earlobes Free earlobes are dominant.

Interlacing fingers Relax and casually fold your hands together so that the fingers interlace. If your left thumb is on top of your right you have the dominant allele *C*. If the right thumb is on top you are homozygous *cc*.

Iris color Brown eyes are dominant over blue.

PTC tasting Touch some test paper impregnated with phenylthiocarbamide to your tongue. If you detect a bitter taste, you have the dominant allele *T*.

Tongue rolling The ability to curl the tongue into a U-shape indicates the dominant allele.

Widow's peak A pointed hairline is determined by the dominant allele.

1. Determine your own phenotype for each trait. Can you determine the genotype, too?

2. The instructor will tabulate data for the whole class for these traits. Are dominant phenotypes always more common than recessive phenotypes?

3. As noted earlier, the Hardy-Weinberg equation works only in ideal cases. However, unless a population is, for some reason, obviously not in equilibrium, the equation will give a reasonable estimate of allele frequencies based on observed phenotypes. Practice using it on the class data for the traits listed previously. Are the recessive alleles more common than you expected? How do the heterozygotes "hide" the recessive alleles in the population?

C. Performing a chi-square analysis

In the following experiment you will examine an actual population of corn kernels and guess what kind of Mendelian ratio it represents. In effect, you will be making a hypothesis about the genotypes of the parents that produced these kernels, so you may find it helpful to try Punnett squares of possible crosses until you discover a ratio similar to what you observe in the population. Then you will use the chi-square test to see if that ratio is, in fact, a probable explanation of what you see.

Chi-square tables are arranged by a value called **degrees of freedom**. This is a statistical term that relates linked phenomena, and it usually equals the number of possible cases minus 1. It is used because linking probabilities often limits the number of outcomes that are possible. As an example, think of a pair of dice. Each die has six sides and when rolled could come up any number between 1 and 6. Now roll both together and total the result. The lowest possible result would be a 2, from rolling two 1s. The highest would be a 12, from two 6s. Thus the result for two dice together could be anything between 2 and 12. This is only eleven possible numbers, even though each die by itself can give six! The degrees of freedom in this case is the original twelve sides of both dice minus one. The missing value is due to the linkage of the two dice. Chi-square tables use degrees of freedom because they examine numbers linked in a ratio.

1. Take an ear of corn from the box marked "Monohybrid Cross." Examine the ear and look for two different phenotypes. **Do not damage the ears – leave the kernels on the cob.** Don't be confused by the phenotypes – this sample only has a few possible test genes. Corn has several color genes but only one will be the test trait on a single monohybrid ear.

2. Count the kernels of the different phenotypes. Make a mark at one end and count all of the kernels in that row. Then go on to the next row and continue until you return to your starting mark.

3. When you are finished, return the corn to the box you took it from.

4. Tabulate your data and find the observed ratio of the phenotypes by dividing the smaller number into the larger and rounding to the nearest whole number ratio.

5. Compare your observed ratio with that obtained by a hypothetical cross:

    Aa x *Aa* gives 3:1
    Aa x *aa* gives 1:1

Choose the closest as your hypothesis – in effect, your hypothesis is an educated guess about the genotypes of the parents of the corn you are examining. The rest of this procedure will test whether your hypothesis is actually close enough to the data to be reasonable.

6. Apply the hypothetical ratio to the total number of kernels you counted to generate the expected number for each phenotype (e.g., if you count 100 kernels and suspect a 3:1 ratio, you would expect exactly 75 of one phenotype and 25 of the other).

7. Calculate the deviation from your hypothesis for each phenotype:

$$\frac{(\text{observed} - \text{expected})^2}{\text{expected}}$$

8. Total these numbers to get a single chi-square value for the entire ratio:

$$\text{chi-square} = \sum \frac{(\text{observed} - \text{expected})^2}{\text{expected}}$$

9. Determine the degrees of freedom for your sample. In this case, the degrees of freedom equals the number of phenotypes minus 1.

10. Consult the row of the chi-square table for this degree of freedom. Find which columns your chi-square total falls between. The top of those columns gives the probability that such a deviation from the expected value would occur by chance. If this probability is greater than 5%, statisticians would say that the hypothesis is consistent with the data. The hypothesis should be rejected if it cannot accurately predict results at least 5% of the time.

11. Now take an ear from the box of "Dihybrid" samples. **These situations are harder to analyze!** Examine the ear for three or four phenotypes. (Dihybrid crosses usually give four phenotypes, but one corn trait, colored vs. colorless, influences a second, purple vs. red, so these ears will only have three phenotypes: purple, red, and colorless.)

12. Count the different phenotypes and estimate the ratio as before.

13. There are several possible dihybrid ratios. The classic 9:3:3:1 ratio only comes from a heterozygous dihybrid cross (i.e., *AaBb* x *AaBb*). Other ratios are possible too (e.g., *AaBb* x *Aabb* gives 3:3:1:1 and *AaBb* x *aabb* gives 1:1:1:1). Moreover, on those where color traits work together to give only three phenotypes, two outcomes are combined (e.g., 9:3:3:1 turns into 12:3:1 or 9:3:4). Use Punnett squares to generate possible hypotheses.

14. Perform a chi-square test as you did for the monohybrid cross.

Sample problem: You count 232 purple and 71 red kernels, a total of 303. This is close to a 3:1 ratio (232/ 71 = 3.3). For an exact 3:1 ratio you would expect 227.25 purple and 75.75 red out of a total of 303 kernels.

Phenotype	Observed	Expected	Chi-square
purple	232	227.25	0.10
red	71	75.75	0.30
Total	303	303.00	0.40

The table indicates that there is a 50 to 80% chance that a deviation from a prediction would occur that shows a chi-square value of 0.40 for one degree of freedom. Since this is greater than 5%, there is no reason to reject the hypothesis that this is a 3:1 ratio from a *Aa* X *Aa*–type cross.

CHI-SQUARE TABLE

Degrees of freedom	Probability of observed outcome						
	0.99	0.80	0.50	0.20	0.10	0.05	0.01
1	0.00016	0.064	0.46	1.6	2.7	3.8	6.6
2	0.20	0.45	1.4	3.2	4.6	6.0	9.2
3	0.12	1.0	2.4	4.6	6.3	7.8	11.3
4	0.30	1.6	3.4	6.0	7.8	9.5	13.3

Deviation insignificant
Hypothesis supported

Deviation significant
Reject hypothesis

IMPORTANT TERMS

allele	dominant	law of independent assortment	Punnett square
antigen	gene	law of segregation	recessive
blood type	genotype	linkage group	Rh factor
chi-square test	Hardy-Weinberg theorem	monohybrid cross	universal donor
codominance	heterozygous	multiple alleles	universal recipient
degrees of freedom	homozygous	pedigree	
dihybrid cross	hybrid	phenotype	

Laboratory Report Genetics

FACTUAL CONTENT

Define the following terms. Brief definitions are given in the chapter, but you may also wish to consult your lecture notes and text for a better idea of how the terms are used. Pay close attention to those words that look similar but have different meanings – you will want to be able to tell them apart if they appear next to each other on a multiple-choice exam.

allele

antigen

blood type

chi-square test

codominance

degrees of freedom

dihybrid cross

dominant

gene

genotype

Hardy-Weinberg theorem

heterozygous

homozygous

hybrid

law of independent assortment

law of segregation

linkage group

monohybrid cross

multiple alleles

pedigree

phenotype

Punnett square

recessive

Rh factor

universal donor

universal recipient

A. Simple dominance, mono- and dihybrid crosses, and Punnett squares

1.   Draw Punnett squares for the following crosses:

   RR x rr

   Rr x Rr

   RR x Rr

   Rr x rr

2.   A woman has type A blood but her father is type O.

   •   What is her genotype?

   •   If she marries a type O man, what proportion of their children would you expect to have type O blood? What proportion should have type A? Type B?

3.   Mary has brown eyes (*BB*). Her father was blue-eyed and her mother was brown-eyed. She marries a blue-eyed man whose parents were both brown-eyed. Mary has a son with blue eyes. Can you figure out the genotypes of everyone in the family? Which ones can you be sure of? Draw a pedigree showing their genotypes. Label the genotypes you are unsure of as "*B?*".

4. *Megacoccus multichromatium* is a strange creature known for its colorful shells and also for its ability to store enormous amounts of sucrose in its body. A scientist obtained two purebred strains of *M. multichromatium*, the first of which was dark brown and always gave dark brown progeny, and the second of which was yellow and always gave yellow progeny. He mated these and obtained offspring that were all dark brown. However, when he crossed these second generation (heterozygous) brown creatures, the 32 offspring were colored as follows:

> 18 dark brown
> 6 red
> 6 light green
> 2 yellow

How many genes (not alleles!) control shell color in *M. multichromatium*? Draw Punnett squares for both the original cross between brown and yellow and the second cross between heterozygotes.

B. Simple dominance in human traits

Complete the table with the numbers of students in your class showing the dominant and recessive phenotypes for each trait. Then calculate the frequency of the recessive phenotype (i.e., the fraction of the population showing that phenotype) and use that value to estimate the frequency of the recessive and dominant alleles.

trait	number showing dominant phenotype	number showing recessive phenotype	frequency of recessive phenotype	frequency of recessive allele	frequency of dominant allele
Acondroplastic dwarfism					
Albinism					
Attached earlobes					
Interlacing fingers					
Iris color					
PTC tasting					
Tongue rolling					
Widow's peak					

Are the recessive alleles more common than you expected?

How do the heterozygotes "hide" the recessive alleles in the population?

C. Performing a chi-square analysis

1. Monohybrid Cross
    List the two different phenotypes on your ear of corn and the number of kernels exhibiting each phenotype:

phenotype	number observed

    Find the observed ratio of the phenotypes by dividing the smaller number into the larger and rounding to the nearest whole number ratio.

    Is your observed ratio closer to a 3:1 ratio or a 1:1 ratio?

    Draw a Punnett square to show what this ratio suggests about the parents' genotypes and how they could have produced these offspring:

What is the total number of kernels on your ear of corn? Based on your hypothetical phenotype ratio, how many kernels of each phenotype would you expect to see for this number?

Calculate the deviation from your hypothesis for each phenotype and total these numbers to get a single chi-square value for the entire ratio:

Consult the chi-square table. What is the probability that the deviation from your hypothesis that you have observed would happen by chance? Is this probability so low that you should reject your hypothesis?

2.   Dihybrid Cross
List the different phenotypes on your ear of corn and the number of kernels exhibiting each phenotype:

phenotype	number observed

Find the observed ratio of the phenotypes by dividing the smaller number into the larger numbers and rounding to the nearest whole number ratio.

Which ratio is closest to your observed ratio — 9:3:3:1, 3:3:1:1, 1:1:1:1, 12:3:1, or 9:3:4?

Draw a Punnett square to show what this ratio suggests about the parents' genotypes and how they could have produced these offspring:

What is the total number of kernels on your ear of corn? Based on your hypothetical phenotype ratio, how many kernels of each phenotype would you expect to see for this number?

Calculate the deviation from your hypothesis for each phenotype and total these numbers to get a single chi-square value for the entire ratio:

Consult the chi-square table. What is the probability that the deviation from your hypothesis that you have observed would happen by chance? Is this probability so low that you should reject your hypothesis?

DNA and Chromosomes

12

INTRODUCTION

In chapter 11 we considered a gene to be a kind of code that determines a particular trait. This is indeed how genes function, but they can also be thought of physically as lengths of DNA. The genetic code is carried in the sequence of nucleotide bases that make up the DNA polymer. Generally, the DNA base sequence is used by the cell as a blueprint for the amino acid sequence of a protein. As such, the DNA code indirectly determines the conformation, and therefore any enzymatic activity or other function, of that protein. Thus DNA ultimately carries the instructions that control the chemistry of the cell. It is through these chemical reactions that visible genetic traits are created. The study of genetics at the molecular level, including how genetic instructions are translated into chemical processes, is called **molecular biology**. Molecular biology is one of the fastest growing fields in biology today.

CHROMOSOMES

As we have seen, most of the DNA of a eukaryotic cell is located in the nucleus. Although mitochondria and chloroplasts also contain tiny amounts of DNA — vestiges of their endosymbiotic past — this extranuclear DNA is only inherited from the mother and does not follow the regular rules of genetics. Therefore in this chapter we will focus on nuclear DNA.

Within the nucleus, the DNA is divided into distinct packets — the chromosomes. Each chromosome is a molecule of DNA, millions of base pairs long, supported by an intricate scaffold of proteins and other molecules. In a nondividing cell, the scaffold is relaxed so that genes can be read as needed. During mitosis the chromosomes contract into compact units (which can be seen with a micro-

125

scope if they are appropriately stained) so that they can be properly sorted amongst the daughter cells.

Long ago geneticists began using fruit flies as experimental organisms because they were easy to raise, easy to cross, and produced the large numbers of offspring needed for statistical studies. As luck would have it, fruit flies also have another useful characteristic: during their larval stages the cells of their salivary glands contain giant **polytene chromosomes**. A polytene chromosome forms when a chromosome replicates as if it were preparing for cell division, but instead of separating in mitosis, the replicated strands stay stuck together. As this process repeats, the chromosome grows into a thick mass made of hundreds of identical copies, all perfectly aligned so that the whole looks like a single magnified chromosome. This unusual phenomenon means that in fruit fly larvae, chromosomes can be viewed with a microscope even in nondividing cells. Polytene chromosomes

structure, not genes), let alone a single base molecule. To study the base sequences we must use chemical methods.

As we discussed in chapters 5 and 6, an individual nucleic acid polymer chain is shaped like a flexible comb, with base molecules sticking out from the polymeric backbone like the teeth of the comb. A molecule of DNA is double-stranded, containing two such "combs." Watson and Crick pointed out that the shapes and polarities of the nucleotide bases were such that the base adenine would have a strong tendency to stick like a magnet to the base thymine, and similarly the base cytosine would stick to guanine. Thus, in DNA, the twin "combs" face each other and stick together "tooth to tooth" to form a ladder shape. It is important to understand that the bases pair because of polar attractions and not permanent chemical bonds. This makes it possible for the two strands to "unzip," exposing the bases so that they can act as a template to replicate the other half. The exposed bases attract the complimentary bases of nucleotide monomers, which, when polymerized into a single chain, become a new matching strand. This is how DNA replicates in the cell prior to mitosis, and it is also the principle behind the complex process of reading DNA and translating it into protein.

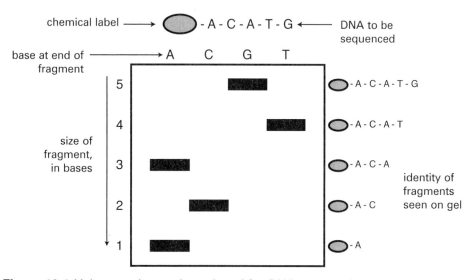

Figure 12.1 *Using an electrophoresis gel for DNA sequencing*

are large enough to reveal many structural details, visible as distinctive banding patterns. This makes chromosomal breaks and abnormalities easier to see. Much of what we know about how chromosome structure influences gene expression has been based on observations of polytene chromosomes.

DNA BASE-PAIRING AND SEQUENCES

When you look at a chromosome under the microscope, you see a complex structure incorporating many macromolecules. It is impossible to see an individual gene (the bands seen in stained polytene chromosomes are density differences in the overall

Base-pairing is a very useful chemical property of DNA. When scientists isolate small quantities of a particular piece of DNA, they can mimic the replication process in a test tube to make multiple copies of the same piece until they have a large enough sample to study. Because this reaction centers on making a new polymer to match the original sample, it is called the **polymerase chain reaction**, usually abbreviated **PCR**. (*Polymerase* refers to the enzyme that actually welds the nucleotides into a single polymer, while the *chain reaction* describes the amplification process.) PCR is frequently used by criminologists to amplify DNA samples taken from crime scenes.

Once enough DNA has been collected, attempts can be made to determine the sequence of bases it contains. In one common method, the DNA is chemically labeled with a radioactive or fluorescent marker and then divided into four samples. Each sample is then subjected to a treatment that breaks the polymeric backbone after a particular base. That is, one sample has cuts after the A bases, one is cut after T, one after C, and the last after G. Conditions are controlled so that none of the samples is completely digested, but instead they contain a mixture that includes strands of different lengths cut after every possible base. Another method uses a base-pairing reaction to make various-length strands from the original sample. Either way, the results end up being four samples, each one a mixture of different-length fragments ending in a particular base.

To find out in what order these bases occur, the samples must then be sorted by size. This is done by electrophoresis. The method is similar to the vertical gel we saw in chapter 6, but since it must separate fragments differing by as little as one bp, it uses very concentrated acrylamide under very high voltage. When the separation is complete, the fragments are detected by virtue of the label incorporated at the beginning of the procedure to give a picture similar to that shown in figure 12.1. At this point the sequence can simply be read from the bottom up. A single gel may reveal a sequence of about 100 to 200 bp. Obviously, many gels must be run to get the sequence of a single gene, which may be thousands of bp long. Laboratories that do a lot of DNA sequencing now have automated methods for speeding up the process.

RESTRICTION ENZYMES AND DNA "FINGERPRINTS"

One of the major discoveries that made genetic engineering possible was the discovery of **restriction endonucleases** (often simply called *restriction enzymes*). These enzymes cut lengths of DNA at specific sequences of bases called **restriction sites**. There are hundreds of known restriction endonucleases, and each enzyme recognizes its own restriction site sequence. Restriction enzymes are the tools that molecular biologists rely on to manipulate genes and other pieces of DNA.

Restriction enzymes are useful because they will only cut DNA at those sequences where their particular restriction sites occur. It is therefore possible to "map" the restriction sites of a given piece of DNA by determining where it is cut by different enzymes. Making a **restriction map** is usually one of the first steps in sequencing a gene – it is a lot easier to sequence the short restriction fragments obtained when a gene is cut up by restriction enzymes than to sequence the entire gene at once. The restriction map can then be used to line up the sequenced fragments to produce the whole sequence. Restriction maps are also used in splicing or altering genes because they indicate points where a piece of DNA can be cut and a new gene inserted.

Restriction maps often show the sites for two or more restriction enzymes. For example, DNA from the bacteriophage lambda virus is about 48,500 base pairs (48.5 **kilobases**, or kb) long. It is cut almost exactly in half by the enzyme Xba I, while Xho I cuts it into two different fragments:

| Xba I | 24.5 | 24.0 |
| Xho I | 33.5 | 15.0 |

(The numbers tell how many kb long each fragment is.)

Digesting lambda DNA with both Xba I and Xho I results in three fragments (24.5 kb, 11.0 kb, and 15.0 kb) since the Xba I site is within the large Xho I fragment and therefore cuts that fragment into two subfragments. Thus this map shows how lambda DNA could be cut into three small pieces as a prelude to sequencing and also indicates two possible sites where the virus could be opened up and a new gene inserted. Because bacteriophages can carry DNA into host bacteria, inserting genes into phage DNA is one way to create new types of bacteria.

Restriction enzymes are also the key to making so-called DNA fingerprints. The DNA in a human cell is estimated to contain over one billion bases, far too many to sequence easily. A restriction map is easier to make than a complete sequence, but even this is impractical for so much DNA. The technique of DNA fingerprinting is similar, but it is designed for characterizing such complex samples.

The actual genes of an individual organism are not as unique as a person's fingerprint – after all, they code for the enzymes that have evolved to be most effective in the species as a whole and cannot show much variation from individual to individual. But one of the odd things about eukaryotic DNA is that much of it does not code for genes. Although the function of this "silent" DNA is unclear, we do know that it tends to mutate easily. As a result, the silent DNA in each individual may be unique. The individual patterns brought out by DNA fingerprinting result from cuts to restriction sites in silent DNA.

As you know from chapter 6, eukaryotic DNA is so complex that it produces countless fragments when digested with a single enzyme, and when electrophoresed and stained, such patterns simply look like smears. Therefore DNA fingerprinting uses radioactive probes that will base-pair with specific DNA sequences, effectively picking out a particular subset of fragments for detection. As an example, say that you have a probe for the DNA coding for a particular enzyme. The probe may only be a few kilobases long, but it will stick to longer fragments if some portion of the longer fragment contains a stretch of bases complimentary to the sequence of the probe. That way a probe for a gene that does not vary much from person to person can still detect sequences that also include stretches of unique silent DNA. If you perform an analysis on two individuals using that probe, you might find that in one individual the probe detects a single fragment 20 kb long, whereas in the second individual the probe detects two fragments 10 kb long, indicating that in the second individual the probed sequence is interrupted by a restriction site not found in the first.

In practice, DNA fingerprinting is a complex procedure, and the results obtained depend greatly on what enzymes and probes are used. Furthermore, no procedure comes close to examining all of the DNA in a particular specimen, so estimates of genetic uniqueness or identity must be based on careful studies of how prevalent particular patterns are in the relevant populations. This can create real problems – for example, one issue in the O. J. Simpson trial was whether enough was known about DNA patterns in African-Americans to make valid estimates of the likelihood that the samples obtained from the crime scene were unique to Simpson. In exercise D you will do a simulation that illustrates how important these factors can be.

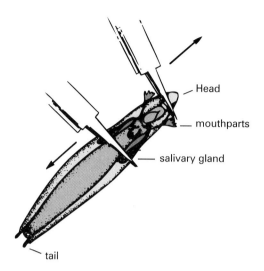

Figure 12.2 *Dissecting the salivary glands from a fruit fly larva*

EXERCISES WITH DNA AND CHROMOSOMES

A. Viewing fruit fly polytene chromosomes

The salivary gland chromosomes of fruit fly larvae are large and impressive, but making a good slide takes skill and practice. Expect to make several slides before you get a preparation that shows the chromosomes clearly. Even after several attempts, not everyone will succeed; if you find that you have made a good slide, be sure to tell your instructor and let your fellow students look at your slide. In an average class there should be enough usable slides for everyone to get a chance to see what the chromosomes look like.

REAGENTS

acetic acid solution	45% acetic acid
aceto-orcein stain	2% orcein in 45% acetic acid

1.   Obtain a vial of fruit fly culture. The adults should have been removed. Examine the medium for little white larvae; the ideal specimens are the large ones crawling up the side of the vial preparing to pupate. Those that are already pupating are not suitable for this exercise – they have immobilized themselves on the glass and are usually darker in color than the active larvae.

2.   Remove a larva and place it on a microscope slide. Place the slide under a dissecting microscope so that you can see what you are doing under magnification. Place a drop of acetic acid solution on the larva. This will irritate the larva, but it is imperative that you keep the preparation moist throughout the procedure.

3.   Locate the head of the larva, the end with the mouthpart appendages. Use fine forceps to grasp the head where the mouthparts originate. Use another pair of forceps to grasp the body of the larva just behind the mouth. Pull the larva apart with the forceps (figure 12.2).

4.   The salivary glands are clear sacs that trail from the head. Search for them under the microscope. There will be a fair amount of white tissue obscuring the salivary glands – tease this away with your forceps. You want to isolate the salivary glands in the center of the slide. Again – be sure to keep the glands moist with acetic acid.

5.   Once you have the salivary glands by themselves, cover them in a drop of stain. Let them sit in the stain for five to ten minutes so the chromosomes can take up the stain.

6.   Place the slide on a paper towel and cover the stained glands with a cover slip. Place another piece of paper towel over the coverslip. Put a rubber stopper on top of the coverslip and press down firmly on the stopper for 20 to 30 seconds.

7.   Remove the slide from the paper towels and place it on the stage of a compound microscope. Look for pink-stained cells and center some of these cells in the field. The chromosomes will not be visible until you switch to high power.

8.   Observe the pattern of bands on the chromosomes. These indicate subtle structural differences along the length of the chromosome and are useful as landmarks for those who wish to map out the details of the chromosomes, but to a beginner such details are overwhelming. Instead, just get an overall sense of what the chromosome looks like. How do these compare with the mitotic chromosomes you saw in chapter 10? Are they longer or shorter? Are the polytene chromosomes well separated from each other? Do you see any areas where the chromosomes appear puffed and diffuse? What might be the reason for this?

T C G A

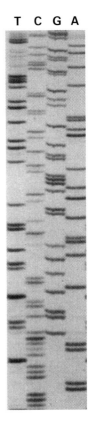

Figure 12.3 *A DNA sequencing gel*

B. Reading a DNA sequence

The chemistry of DNA sequencing is complicated and is best left to experienced professionals, but the overall process is similar to the procedures you used in chapter 6. Figure 12.3 is a photograph of a finished sequence gel, which you can examine for data.

1.   Locate the lowest band on the gel. What base column is this in? That is the first base of the sequence.

2.   Move up the gel to the second lowest band. What base does this represent?

3.   Continue working your way up the gel. It is easier if you hold a ruler under the band you are working on and move this up as you go. It is also easier if you work with a partner, one person reading while the other records the sequence. When you are finished, the partners can switch roles and repeat the process to proofread their results.

4.   You should now have a string of letters in your notes showing the DNA code from this strand. Opposite each letter write the letter of the complimentary base that would be paired with that base in the double helix.

5.   How many bases have you read? It takes three bases to code for a single amino acid — how many amino acids could your sequence code for? Assuming an average molecular weight of about 100 Daltons per amino acid, what would be the molecular weight of the protein coded for by this DNA? Refer to your notes from chapter 6. How does the molecular weight of the proteins you examined then compare to that of the protein encoded by this DNA?

C. Analyzing a restriction map

Again, in this exercise you will not be asked to carry out a digestion or run a gel. Figure 12.4 illustrates an electrophoresis gel that has three samples of a small animal virus, adenovirus-2, digested with different restriction enzymes. The intact viral DNA is 35.9 kb long. Cutting adenovirus-2 with Bam H I generates four restriction fragments ranging in size from 4.7 kb to 14.3 kb. Digestion with Cla I releases two large fragments (17.7 kb and 17.3 kb) and a small one (0.9 kb). When digested with both enzymes, the result is six fragments of various lengths. All three digests were electrophoresed next to each other. We have already calibrated the gel and calculated the size of the fragments for you. The size of each fragment, in kb, is indicated beside the gel. Your task is to analyze the gel data and construct a restriction map based on this data. Assume that through chemical labels you already know that the ends of the viral DNA are on the 10.7 and 14.3 kb Bam H I fragments, the 0.9 and 17.3 kb Cla I fragments, and the 0.9 and 14.3 kb fragments from the dual digest. By convention, the 10.7 kb Bam H I fragment of adenovirus-2 is considered to be on the left side of the map. You are to study the overlaps in the digest and order the remaining Bam H I fragments and all of the Cla I fragments to make a map of the restriction sites for these two enzymes. Write the Cla I map under the Bam H I map so that it looks similar to the Xba I / Xho I map for phage lambda DNA in the example discussed earlier.

1.   Start by listing the length of each fragment from each digest in your notes.

2.   Make a preliminary map for Bam H I, since you know the most about it: 10.7 is on the left and 14.3 is on the right. In between are the 4.7 and 6.2 kb fragments, but since you do not know what order they are in, leave the middle blank for now. You may find it helpful to draw the map to scale on your graph paper (e.g., 1 square = 1 kb).

3.   Below the Bam H I map start a Cla I map. One of the end Cla I pieces is cut by Bam H I. Use the dual digest data to figure out which one. This will allow you to order all three Cla I fragments.

4.   Now go back and use the dual digest data and the Cla I map to complete the Bam H I map. Record the entire map in your notes.

D. DNA fingerprinting

In the following simulation you will explore how enzyme digests can produce different patterns, and by comparing your results with those of your classmates under different conditions, you will examine what factors influence whether the patterns obtained are unique.

1.   Fill in the boxes in figure 12.5 with a random sequence of nucleotides (i.e., A, T, C, or G) to represent the variable DNA. The other sequences are conserved DNA and will represent the areas that are probed.

2.   Your instructor will describe the particular sequences cut by some restriction enzymes. Examine your sequence for these sites and mark where your DNA would be cut.

3.   Construct a chart to show how an electrophoresis gel of your fingerprint would look. On the left, make a vertical axis marking off fragment lengths from 0 to 100 bases. Along the top, mark lanes (columns) representing a sample digested with each enzyme or group of enzymes.

4.   For each lane, note how many fragments would be produced by the enzyme or combination of enzymes, and how many bases long each fragment would be. Since the probes will only detect the constant regions, count only those fragments that contain at least six bases from one of the constant regions.

5.   Mark the fragments in the appropriate lanes according to their size. If two or more fragments are the same size, simply make one mark for all of them – a gel could not resolve such fragments.

6.   The pattern on the entire "gel" is your DNA fingerprint. Compare yours to those of your classmates. Is your pattern unique? How would varying the enzymes and their combinations alter the uniqueness? How would changing the probes alter the uniqueness? Do you think a larger DNA sample would make a difference?

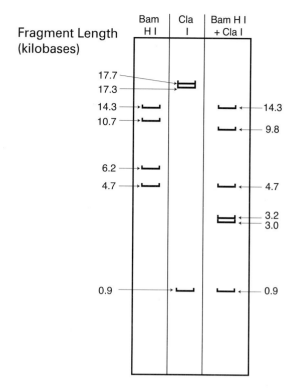

Figure 12.4 *Restriction digest of adenovirus-2*

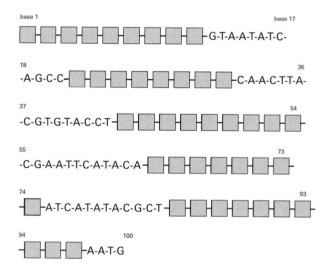

Figure 12.5 *Sample for DNA fingerprint simulation*

IMPORTANT TERMS

kilobase (kb)
molecular biology

polymerase chain reaction (PCR)
polytene chromosome

restriction endonuclease
restriction map
restriction site

Laboratory Report DNA and Chromosomes

FACTUAL CONTENT

Define the following terms. Brief definitions are given in the chapter, but you may also wish to consult your lecture notes and text for a better idea of how the terms are used. Pay close attention to those words that look similar but have different meanings – you will want to be able to tell them apart if they appear next to each other on a multiple-choice exam.

kilobase (kb)   restriction endonuclease

molecular biology   restriction map

polymerase chain reaction (PCR)   restriction site

polytene chromosome

EXERCISES - Record your data and answer the questions below.

A. Viewing fruit fly polytene chromosomes

Make a sketch of the chromosomes as they appear under high power:

How do these compare with the mitotic chromosomes you saw in chapter 10? Are they longer or shorter?

Are the polytene chromosomes well-separated from each other?

Do you see any areas where the chromosomes appear puffed and diffuse? What might be the reason for this?

B. Reading a DNA sequence

Record the sequence of DNA bases read from the gel. Next to each base, write the letter of the complimentary base that would pair with it in a complete double helix:

base	complimentary base	base	complimentary base	base	complimentary base

How many bases have you read? How many amino acids could your sequence code for?

Assuming an average molecular weight of about 100 Daltons per amino acid, what would be the molecular weight of the protein coded for by this DNA? Refer to your notes from chapter 6. How does the molecular weight of the proteins you examined then compare to that of the protein encoded by this DNA?

C. Analyzing a restriction map

List the length of each fragment from each digest:

Bam H I	Cla I	dual digest

Work out the restriction map on the graph paper:

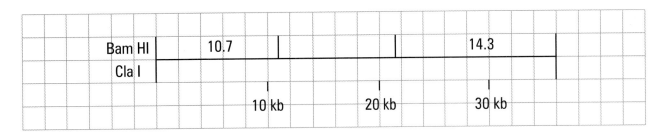

D. DNA fingerprinting

List the restriction enzymes and the sequences of their restriction sites that you will be using for this exercise:

Construct a chart to show how an electrophoresis gel of your fingerprint would look. Along the top mark lanes (columns) representing a sample digested with each enzyme or group of enzymes.

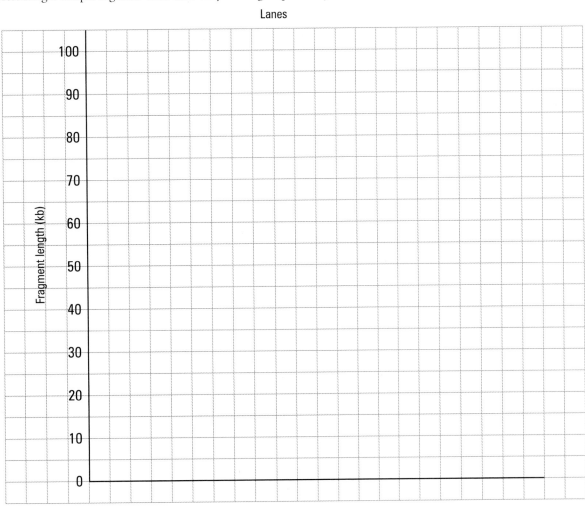

Compare your "DNA fingerprint" to your classmates. Is your pattern unique?

How would varying the enzymes and their combinations alter the uniqueness?

How would changing the probes alter the uniqueness?

Do you think a larger DNA sample would make a difference?

Variation in Populations

13

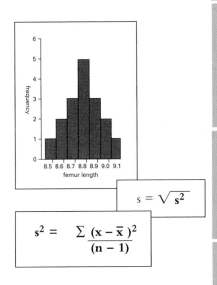

$$s = \sqrt{s^2}$$

$$s^2 = \frac{\sum (x - \bar{x})^2}{(n - 1)}$$

OBJECTIVES

New Skills

In this chapter you will learn how
1. to describe the variation in a population as a frequency distribution.
2. to calculate important characteristics of a frequency distribution: the mean, variance, standard deviation, and coefficient of variation.
3. to graph a continuous frequency distribution as a histogram.
4. to plate out bacteria in order to study their response to specialized media.

Factual Content

In this chapter you will learn
1. some basic terminology relating to variation in populations, the statistics of frequency distributions, and natural selection.
2. the statistical formulas for calculating the mean, variance, standard deviation, and coefficient of variation of a distribution.

New Concepts

In doing the exercises in this chapter, you will observe
1. phenotypic variation within populations and how it can be summarized statistically.
2. how a frequency distribution can shift over time, changing the character of the population.
3. how specific selective forces acting on a population, for example, the presence of an antibiotic among bacteria, can produce a change in the population over time.

INTRODUCTION

When we considered genetics earlier, we concentrated on individual organisms and how an allele or combination of alleles in each organism might be expressed. Because these principles apply to every individual in a population, the study of population genetics can be quite complex. One of the more important aspects of a population is how much variation it holds for various traits. An endangered species, for example, may have so few survivors that they may not have the genetic variation necessary to allow the species to adapt to new challenges. On the other hand, a populous species like humans may include so much variation that determining standards of "normal health" is difficult, leading to occasional unpredictable responses to drugs or diseases. When the genetics of a trait are well understood and the various alleles identified, it may be possible to develop a very detailed genetic description of that trait throughout a population. Even when this information is lacking, however, the extent of phenotypic variation within a population can still be sampled and measured. The patterns of variation so elucidated may tell us a great deal about the population, its history, and its ability to change in the future.

TYPES OF VARIATION

The traits described by classic Mendelian genetics can only exhibit one of several alternative states, without any intermediate states. When such a trait has only two alleles, then everyone in the population must exhibit either one phenotype or the other. In human blood types, for example, everyone is either Rh positive or Rh negative – there is no in-between factor. Multiple alleles and phenomena such as codominance can complicate the picture, as we know from studying the ABO system of blood types, but nevertheless the possibilities remain limited. Traits that have only a few discrete phenotypic states are called **meristic traits**. We shall consider meristic traits in more detail in chapter 14.

Most of the visible variation among individuals in a population is not due to obvious meristic traits. In humans, traits such as height and skin color have a wide range of possible values with a full continuum of possible intermediate states. Characteristics that can be measured on a continuous scale are called **continuous traits**. Such traits are not determined by a single gene but by several genes interacting with each other and with the environment. When describing the possible phenotypes of a continuous trait we are forced to generalize and round off. For example, a person may be described as being 70 inches tall, but this might actually represent a measured height anywhere from 69.5 to 70.49 inches, rounded off to the nearest inch. By rounding off we avoid unnecessary details, although by doing so we group together a number of people who are not *exactly* the same height. The degree to which we round off depends on how we intend to use our data: continuous traits are best described statistically, and the extent of rounding off determines how much detail will be included in the statistical summary of the population.

QUANTIFYING VARIATION

Long ago, mathematicians working with statistics developed the idea of a **frequency distribution** to express and measure the variation within a sample of individuals. In biology, a frequency distribution is a representation of the number of individuals in each phenotypic class present in the sample. For a simple meristic trait with few phenotypes, a frequency distribution can be represented by a table listing how many specimens exhibit each phenotype. This method will not do for continuous traits, there being too many possible phenotypes to tabulate. Instead we use statistical methods to summarize all of the data with a few descriptive terms.

These methods are best illustrated by applying them to a hypothetical population. Imagine that a paleontologist is trying to

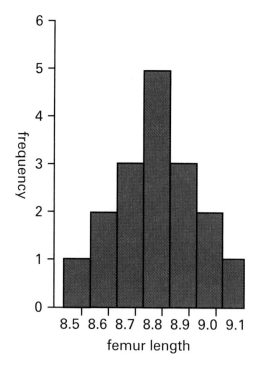

Figure 13.1 *Frequency of femur lengths in fossil sample*

compose a description of an extinct species of animal. She has a sample of femurs (thigh bones) and measures their length in order to establish the "normal" size for the creature. She obtains the following values (in cm): 8.8, 9.0, 8.7, 8.9, 8.9, 9.1, 8.5, 8.8, 8.7, 8.6, 9.0, 8.8, 8.6, 8.8, 8.9, 8.7, 8.8. It is difficult to discern any pattern in variation by simply looking at the raw data. A better presentation of the data is in the form of a kind of bar graph called a **histogram** (figure 13.1). In a histogram, the value of the trait is indicated on the x-axis and the number of times it is observed is indicated on the y-axis. The bars of a histogram are always shown touching along the side in order to remind the viewer that the real values are continuous, even though they are rounded off.

In this example the frequency distribution shows a strong bias toward values in the middle of the range: many individuals are mid-sized with fewer and fewer individuals at the extremes. As a result, the histogram approximates a "bell-shaped" or **normal curve**. Many biological traits have a normal distribution.

As graphic representations of frequency distributions, histograms are very effective, but they are clumsy tools for comparing several different distributions. This is where it is useful to have a numerical index to summarize the entire distribution. We will consider three statistics that together give a very good description of a frequency distribution.

The center of the distribution is the **mean**. The mean is simply the arithmetic average of the data:

$$\bar{x} = \sum x/n$$

where $\bar{x}$ is the mean, x are the individual data values, and n is the number of data values.

Variance describes how the data spread out around the mean. The lower the variance, the more similar the individuals are to one another. Variance is calculated by the formula:

$$s^2 = \frac{\sum (x - \bar{x})^2}{(n - 1)}$$

where s^2 is the variance and the other variables are as above. The term $(x - \bar{x})$ indicates how much any individual value deviates from the mean. Thus the variance represents the average *squared* deviation for the sample. This may seem odd, as an average is usually obtained by dividing a total by n instead of $n - 1$. Variance is unusual because it is derived from another statistic, the mean, and as you learned in chapter 11, when two statistics are linked, we must account for the linkage by using degrees of freedom. The average squared deviation is used instead of the average deviation, because the average deviation always equals zero – the highs and the lows cancel each other exactly (try it on this example yourself and see).

The **standard deviation** of a distribution is more descriptive for many applications than the variance. The standard deviation is simply the square root of the variance:

$$s = \sqrt{s^2}$$

In order to obtain the standard deviation, we must first square the deviation (to get the variance) and then take the square root. This may seem like unnecessary work, but it is not: as noted, the deviation must be squared before taking the average or the the the average will be zero, and taking the square root after averaging returns us to the original units of the measurement. In our example of femur lengths, both the average and the standard deviation would be in centimeters.

Table 13.1 shows how these statistics would be derived from our hypothetical example. Many calculators will do these calculations automatically, but it is important that you understand how the process works.

NATURAL SELECTION AND SHIFTS IN FREQUENCY DISTRIBUTIONS

Frequency distributions are very useful for studying the characteristics of a population over time. Under ideal conditions, a population in genetic equilibrium will behave according to the Hardy-Weinberg theorem: alleles will be shuffled about randomly, but their overall frequencies will not change. Obviously, in such a situation, genetically based variation will not change either, and the frequency distributions of those traits will be stable.

Yet Hardy and Weinberg knew they were dealing with a theoretical abstraction. By the time they came up with their theorem it had long been appreciated that it is quite normal for populations to change. In the middle of the eighteenth century, Benjamin Franklin wrote many articles arguing that Great Britain should give its American colonies more autonomy. One of his arguments was that because America was largely undeveloped, it offered enormous potential for new ways of life to thrive. In addition to his other activities, Franklin was one of the foremost scientists of his day, and he was well acquainted with many species of plants and animals that had been introduced into America and had begun to adapt to the new environment. Franklin used this as a metaphor for what the American economy could do if free from interference from Britain.

STATISTICAL ANALYSIS OF FEMUR LENGTH DATA		
Observed length x (cm)	Deviation from mean $x - \bar{x}$ (cm)	Squared deviation $(x - \bar{x})^2$
8.8	0.0	0.00
9.0	0.2	0.04
8.7	-0.1	0.01
8.9	0.1	0.01
8.9	0.1	0.01
9.1	0.3	0.09
8.5	-0.3	0.09
8.8	0.0	0.00
8.7	-0.1	0.01
8.6	-0.2	0.04
9.0	0.2	0.04
8.8	0.0	0.00
8.6	-0.2	0.04
8.8	0.0	0.00
8.9	0.1	0.01
8.7	-0.1	0.01
8.8	0.0	0.00
$\sum x = 149.6$		$\sum (x - \bar{x})^2 = 0.40$

mean: $\bar{x} = \sum x/n = 149.6/17 = $ **8.8 cm**
variance: $s^2 = \sum (x - \bar{x})^2/(n-1) = 0.40/16 = $ **0.025**
standard deviation: $s = \sqrt{s^2} = \sqrt{0.025} = $ **0.16 cm**

Table 13.1

As we all know, the British did not take Franklin's advice, but his ideas did circulate in England, where, appropriately enough, they were adapted to local conditions. In the north of England the industrial revolution was in full swing, sparking a population boom in many cities. But here the land was long settled, and overcrowding was the result. Thomas Malthus, a clergyman interested in economics, took note of this. He was more pessimistic than Franklin, and he saw population growth as a recipe for suffering rather than opportunity. As he wrote in his *Essay On The Principle Of Population,*

> . . . nature has scattered the seeds of life abroad with the most profuse and liberal hand [but] has been comparatively sparing in the room and the nourishment necessary to rear them.

Malthus believed that the poor were destined to suffer, for over-population must lead inevitably to famine, disease, and privation. These factors would then help to check further population growth.

Malthus's essay was widely read. In the middle of the nineteenth century, Charles Darwin remembered it as he contemplated natural biodiversity. Darwin was a professional naturalist and had been sent to South America to catalog as many species of plants and animals as he could. In the course of his work he developed a keen appreciation for the amount of phenotypic variation that existed in wild populations. Darwin realized that if species produced more offspring than resources could support (as Malthus contended), then those individuals in the species who, by natural variation, had some kind of advantage over the others would be most likely to survive and reproduce. Their variations would gradually become more common. He wrote that

> . . . natural selection is daily and hourly scrutinizing, throughout the world, the slightest variations, rejecting those that are bad, preserving and adding up all that are good; silently and insensibly working . . . at the improvment of each organic being . . . .

The principle that differing levels of fitness determine which types of organisms survive and reproduce is called **natural selection**. Others before Darwin had suggested theories of evolution, but these tended to be based on the idea that characteristics acquired during life could be passed on to offspring if they proved useful. Darwin realized that novel characteristics need not be acquired during life because they already exist within the natural variation among the different members of a population. Hence, biologists interested in evolution spend a great deal of time studying variation.

A common misconception about evolution is that it involves changes to an individual that are then somehow passed on to future generations. *The fundamental unit of evolution is not the individual, but the population.* The population changes from generation to generation because the frequency distributions of the traits within the population shift. A classic example of how this works is the story of the peppered moth (*Biston betularia*), which exhibits both light- and dark-colored phenotypes. Two centuries ago the lighter moths were the more common, but in areas that have been exposed to industrial pollution and soot, the darker variety has become more numerous, as birds have a hard time finding and eating the dark moths when they alight on sooty surfaces. Natural selection did not cause the light moths to become dark, it caused the dark coloration to become more frequent over several generations by favoring the dark individuals. The evolutionary change is described by the shifting frequency distributions of the light and dark phenotypes. All evolutionary changes are essentially shifts in frequency distributions.

EXERCISES WITH CONTINUOUS VARIATION

A. Measuring continuous variation in beans

Variation is a characteristic of populations. Any study of variation must, of necessity, involve a great deal of counting and repetitive measurements. These measurements may seem tedious, but they are the only way to obtain a representative sample of an entire population. When an anthropologist looks at a box of fossilized teeth and determines that some are from *Australopithicus africanus* and others are from *Australopithicus afarensis*, he can do so only because he is thoroughly familiar with the traits of each species.

1. You will be given samples of two kinds of beans (e.g., pinto beans and lima beans). Make a table with two columns so that you can record the length and width of each bean. Then measure each bean with calipers and record the data. Be sure to record the length and width of each bean – you will need to know which length goes with which width in step 4.

2. Following the example earlier in this chapter, calculate the mean, variance, and standard deviation for each dimension of each type of bean. Draw histograms for the length and width of each type of bean, rounding off to the nearest mm.

3. How do the standard deviations for the data from the measurements compare? Is the magnitude of the standard deviation related to the size of the associated mean? Calculate the **coefficient of variance** to correct the standard deviations for the size of the means:

$$V = s / \bar{x}$$

4. Construct a two-dimensional frequency distribution by graphing the length versus the width of each bean. Plot both types of beans on the same graph, using different colors or symbols for the two kinds of beans. Does the variability (indicated by the size of the area of the graph in which data points appear) grow larger as the data values grow larger? That is, are the points for the large beans more spread out than those for the small beans?

B. Modeling a selective shift in frequency distribution

To study evolution in a population you would have to gather measurements like those in exercise A for several generations to see what, if any, changes occurred. In this exercise you will instead use your data to create a simulation of what might happen if your population of beans were subjected to selective forces that favored the largest individuals. This type of selection has been used throughout history to breed strains of domestic plants and animals with economically important traits.

1. Assume that you are trying to breed a bigger bean and will only plant those beans that are larger than the mean length.

2. Assume further that the average length of the next generation will be one standard deviation larger than the population you started with and that the coefficient of variation will be the same (this means that the standard deviation will be bigger). Calculate what this would be for generation 2.

3. Repeat the process generation after generation, until you get a population whose mean length is at least twice that of the original population. How many generations did this change take?

Example:

33 beans are measured.

$$\bar{x} = 12.5 \text{ mm}$$
$$s = 1.1 \text{ mm}$$
$$V = 0.09 \ (\text{i.e., } 9\%)$$

Generation 2 would be:
$$\bar{x}_2 = 12.5 + 1.1 = 13.6$$
$$s_2 = 13.6 \times 0.09 = 1.2$$

Generation 3 would be:
$$\bar{x}_3 = 13.6 + 1.2 = 14.8$$
$$s_3 = 14.8 \times 0.09 = 1.3$$

And so on until x_n is at least 25 mm.

C. Variation and selection in bacteria

The study of variation is facilitated by distinguishing continuous from meristic traits because this helps us focus our methodology. Nature does not always cooperate, however. Some traits have aspects that are both meristic and continuous. Antibiotic resistance in bacteria is such a trait.

Antibiotics are compounds that selectively kill or inhibit the growth of certain kinds of organisms (such as bacteria) while leaving others unharmed. Usually they work by inhibiting a key enzyme in the target species. If the antibiotic inhibits the growth of the species, the species is considered to be sensitive to the antibiotic, and if not, the species is resistant. In this respect resistance is a meristic trait. In some cases, however, there are degrees of resistance – that is, a species may survive low doses of the antibiotic but be sensitive to higher doses. In these situations the variation in resistance may be continuous around a mean dose response.

In this exercise you will subject a population of bacteria to the antibiotic nalidixic acid. To keep the exercise simple, you will only be asked to test two concentrations of the antibiotic, but in principle you could test a wide range of concentrations. You will also see how selection can enrich the population with resistant individuals. As with the work you did on bacterial fermentation in chapter 8, you will have to use sterile techniques to obtain reproducible results.

REAGENTS

nutrient agar plates	petri dishes filled with sterile bacterial growth medium
low-antibiotic NA plates	nutrient agar plates with 20 μg/ml nalidixic acid
high-antibiotic NA plates	nutrient agar plates with 30 μg/ml nalidixic acid
nutrient broth	sterile liquid bacterial growth medium
bacterial culture	pure cultures of *Escherichia coli*, 24 hours old

Period One

1. Your instructor will concentrate a culture of *E. coli* by centrifugation and resuspend the pelleted cells in one-tenth their original volume of nutrient broth.

2. Remove the cap from the tube of concentrated bacteria and pass the mouth of the tube through the flame of a bunsen burner. Draw up 0.1 ml of culture with a sterile pipet as demonstrated by

the instructor and release it onto the surface of a nutrient agar plate.

3.   Dip a glass "hockey stick" in a dish of ethanol and then carefully ignite the stick in the flame of the burner to sterilize it. Use the sterilized stick to spread the bacteria evenly over the surface of the plate.

4.   Repeat steps 2 and 3 for a low-antibiotic plate and a high-antibiotic plate. Incubate all plates as directed.

Period Two

1.   Examine the plain nutrient agar plate for colonies of bacteria. This is an indication of how many bacteria were in the original sample.

2.   Look for colonies in the antibiotic plates. How do the numbers compare to the plate without antibiotics? Use the numbers of colonies to calculate what percentage of the original bacteria were resistant to low levels of nalidixic acid and what percentage were resistant to the higher level.

3.   Flame an inoculating loop to red hot and use the sterile loop to take up a colony of bacteria from the low-antibiotic plate. Transfer the bacteria to a sterile tube of nutrient broth and incubate in a 37° water bath for one hour.

4.   Concentrate the cultured bacteria by centrifugation and resuspend in one-tenth the original volume. Use this culture to plate out a new nutrient agar plate and a new high-antibiotic plate (repeating steps 1 to 3 from period one). Incubate the plates as directed.

Period Three

1.   Examine the plates for colonies. How does the percentage of high-resistance bacteria in this sample compare to the percentage of high-resistance bacteria in the original sample? Do you see evidence of a shifting frequency distribution?

IMPORTANT TERMS

antibiotic	frequency distribution	meristic trait	standard deviation
coefficient of variance	histogram	natural selection	variance
continuous trait	mean	normal curve	

Laboratory Report Variation in Populations

FACTUAL CONTENT

Define the following terms. Brief definitions are given in the chapter, but you may also wish to consult your lecture notes and text for a better idea of how the terms are used. Pay close attention to those words that look similar but have different meanings – you will want to be able to tell them apart if they appear next to each other on a multiple-choice exam.

antibiotic

meristic trait

coefficient of variance

natural selection

continuous trait

normal curve

frequency distribution

standard deviation

histogram

variance

mean

EXERCISES - Record your data and answer the questions below.

A. Measuring continuous variation in beans

Record the length and width of each bean in your samples and then calculate the mean, variance, and standard deviation for each dimension of each type of bean:

length	width	length	width

mean:

variance:

standard deviation:

Draw histograms for the length and width of each type of bean:

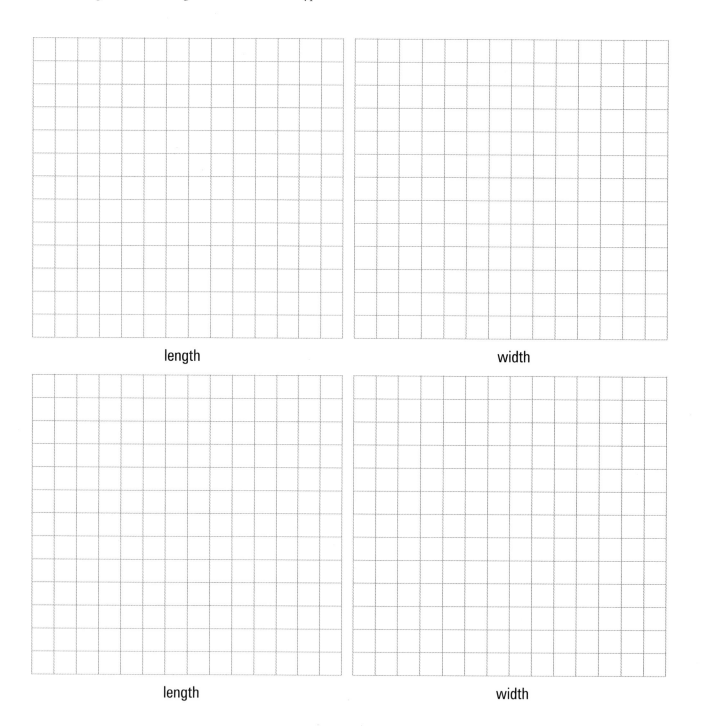

length   width

length   width

How do the standard deviations for the data from the measurements compare? Is the magnitude of the standard deviation related to the size of the associated mean?

Calculate the coefficient of variance for the size of each mean:

Construct a two-dimensional frequency distribution by graphing the length versus the width of each bean:

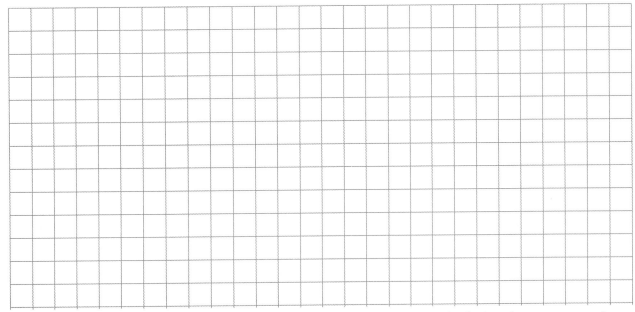

Does the variability grow larger as the data values grow larger? That is, are the points for the large beans more spread out than those for the small beans?

B. Modeling a selective shift in frequency distribution

Record the mean, standard deviation, and coefficient of variation for the lengths of the population of beans you are using for this exercise:

mean:

standard deviation:

coefficient of variation:

Calculate the mean and standard deviation for each generation of selection:

generation	mean length	standard deviation

How many generations does it take to double the length?

C. Variation and selection in bacteria

Period Two
Record the number of colonies on each plate:

plain nutrient agar

low antibiotic

high antibiotic

Calculate what percentage of the original bacteria were resistant to low levels of nalidixic acid and what percentage were resistant to the higher level:

Period Three
Record the number of colonies on each plate:

plain nutrient agar

high antibiotic

Calculate what percentage of this population of bacteria were resistant to high levels of nalidixic acid:

How does the percentage of high resistance bacteria in this sample compare to the percentage of high resistance bacteria in the original sample? Do you see evidence of a shifting frequency distribution?

Variation in Meristic Traits

14

OBJECTIVES

New Skills

In this chapter you will learn how
1. to visualize and classify fingerprints as examples of meristic traits.
2. to anesthetize and handle fruit flies for use in population studies.
3. to recognize some meristic mutant phenotypes in fruit flies.
4. to recognize male and female fruit flies.

Factual Content

In this chapter you will learn some basic terminology relating to fruit fly genetics.

New Concepts

In doing the exercises in this chapter, you will
1. sample some populations in order to observe meristic traits.
2. see how meristic traits exhibit discontinuous frequency distributions.
3. see how meristic traits can be sampled in a population over several generations to examine whether the population is in equilibrium or is evolving.
4. observe how a population can change over time as a result of the reshuffling of alleles in the population and a shift in their relative frequencies.

INTRODUCTION

In many ways, meristic traits are easier to work with than continuous traits: limiting the number of phenotypes makes it simpler to determine quickly what phenotype is exhibited and usually eliminates the need to calculate means and standard deviations. Because of this, meristic traits are ideal for lengthy studies of evolution.

As noted in the previous chapter, **evolution** is visible as a change in phenotype frequency over several generations. Because a trait must have a genetic basis if it is to evolve, evolution must involve a shift in genotype frequencies. Sometimes nonheritable traits change over time: after World War II many human populations all over the world began to get taller, but this was due to changes in nutrition occasioned by the spread of new farming methods. This was not an evolutionary change because it had no genetic basis.

Any genetic trait that shows some variation can potentially be monitored for evolutionary changes. Fruit flies are ideal specimens for such studies because they have brief generations, and much is known about their genetics.

DROSOPHILA MELANOGASTER

Mendel did his work on pea genetics at about the same time as the American Civil War. Darwin published his theory of natural selection during this same period, but Darwin seems to have been totally unaware of Mendel's findings. Indeed, Mendel's work was virtually ignored until the end of the century, when much of it was rediscovered by a new generation of scientists.

One of the most famous of these new geneticists was T. H. Morgan of New York. Instead of peas, Morgan performed his experiments on the common fruit fly, **Drosophila melanogaster.** To this day peas and *Drosophila* remain among the most popular organisms for genetics studies. Fruit flies are in some respects superior to peas for this work in that they can be raised in great numbers in relatively small spaces and they can complete a new generation in two to three weeks. Since the time of Morgan, numerous traits have been examined in *Drosophila*, and many genes have been traced to specific locations on one of the fruit fly's four pairs of chromosomes.

Unlike peas, fruit flies are not domesticated, and Morgan could not start with well-established strains. Morgan and his coworkers identified several meristic traits in fruit flies by comparing rare and unusual individuals with "normal" or **wild**-type flies. Wild flies have red eyes, are free of malformed wings and appendages, and thrive under standard culture conditions. **Mutant** flies stand out by differing from wild in some structural or physiological trait.

The nomenclature of fruit fly genetics has been adapted slightly so that it expresses not only whether a trait is dominant or recessive, but also whether it is a wild or mutant allele. All wild alleles, whether dominant or recessive, are designated by the sign "+." Mutant alleles are given a one-or two-letter abbreviation that is capitalized in the case of dominant alleles. For example, the recessive allele *e*, when homozygous, gives the fly body a dark ebony color. Ebony flies are genotype *e/e*, wild are +/+, and normal-looking heterozygotes are +/*e*. Similarly, plum-colored eyes are caused by the dominant allele *Pm*. Wild-type flies have the genotype +/+, and the plum phenotype can be caused by both the *Pm/Pm* and *Pm/+* genotypes.

DESCRIBING MERISTIC VARIATION

Constructing a frequency distribution to describe a meristic trait requires counting the various phenotypes present, the same as with a continuous trait. In the case of a meristic trait there are fewer possibilities, and the distribution can often be represented as a simple table. For example, the distribution of body color in a hypothetical population of fruit flies could be summarized:

Light body (wild)	Ebony body	Total
74	26	100

This table shows a variation in a single trait only. A two-dimensional table would be needed to describe simultaneous variation in two traits. Assume, for instance, that this fly population also contained a mutant wing shape; the distribution might then look like:

	Light body	Ebony body	Wing shape total
Straight wing	66 (wild)	24	90
Vestigial wing	8	2	10
Body color total	74	26	100

A meristic distribution may also be illustrated graphically. If represented by a bar graph the result is not a histogram, because the distribution is not continuous (so the bars are not shown touching and cannot resemble a normal curve). Figure 14.1 shows two types of graphs that could be drawn to represent the data in the preceding table.

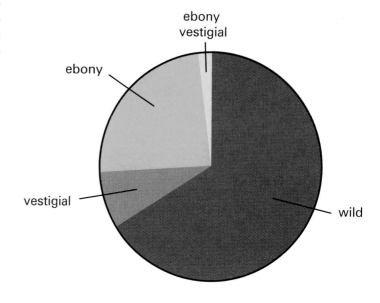

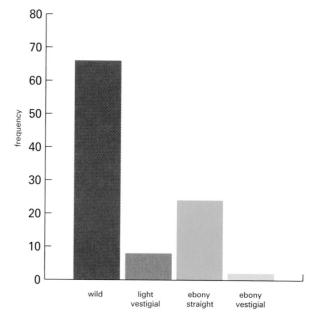

Figure 14.1 *Graphic representations of* Drosophila *phenotypes*

SHIFTS IN MERISTIC VARIATION

Meristic traits can be very important in evolution. Most populations have an assortment of alleles in circulation that have the potential of becoming more frequent should conditions begin to favor them. For example, the hemoglobin allele that is responsible for sickle cell anemia is normally rare, because people who are homozygous for that allele suffer from sickle cell anemia. In regions where malaria is common, however, the sickle cell allele is also common, for heterozygotes with a single sickle cell allele have some resistance to malaria.

Meristic traits can evolve in complex interactions, as in cases involving proteins and the molecules they bind to. Consider the recent discovery of a physiological messenger system governing appetite and obesity in mammals: obese mice were found to be genetically unable to produce the chemical message that signaled to their brains that it was time to stop eating. These mice lose weight if they are given doses of the chemical message as a drug. When obese humans were studied, the same system was discovered, but with an unfortunate twist – in humans the problem is in the receptor, not the signal molecule. No amount of signal-drug would help affected people lose weight because their receptors cannot respond to any of it. In evolutionary terms, the two traits, signal and receptor, must change together if they are to produce an effective system.

EXERCISES WITH MERISTIC VARIATION

A. Tabulating meristic variation in human fingerprints

Human fingerprints are complex characteristics determined by a multitude of genetic and developmental influences. If the tiny details that make each print unique are ignored, it is possible to classify any fingerprint as one of three possible patterns: arches, loops, and whorls. Even these broad groups do not represent particular genetic alleles, as an individual may exhibit all three patterns on one hand. Nevertheless there are genetic predispositions toward one pattern or another.

The rarest pattern is the arch. It is also the simplest. Loops are the most common. Most loops are described as *ulnar* – that is, the base of the loop opens toward the little finger of the hand (toward the ulna bone of the forearm). If the loop opens toward the thumb (and the radius bone of the forearm), it is called *radial*. Whorls are circular and are more common than arches but not as common as loops.

In this exercise you will classify the fingerprints of each of your fingers, and the instructor will total up the distribution of each pattern for the entire class. We will then examine the data in relation to a second meristic trait, the sex of the individual reporting each type of pattern.

1. Color an area on a piece of scratch paper darkly with a #2 pencil.

2. Examine your fingertip and look for the center of the pattern as illustrated in figure 14.2. Rub that part of your fingertip on your colored scratch paper to darken the fingerprint.

3. Take a piece of clear tape and place it on your fingerprint to pick up the graphite image. Transfer the tape to an index card and label which finger it came from. Repeat for each finger.

4. Classify each print as arch, ulnar loop, radial loop, or whorl. You may need to look at the prints with a dissecting microscope to see them well.

5. Your lab instructor will tally the distribution of each pattern, by sex, for the whole class in a two-dimensional table. Are the patterns distributed differently among the sexes?

6. Construct a bar graph of the data. How does this bar graph compare to the histograms you made in chapter 13?

arch

loop

whorl

Figure 14.2 *Fingerprint types*

B. Studying shifts in meristic variation over time

Evolution is a slow and gradual process. With patience, it can be observed in nature as populations change from generation to generation. Obviously such studies are time consuming, but many have been published. Analogous studies can be performed in the laboratory with fruit flies, which, even in a period as brief as a semester, can produce four to six generations.

It is important to remember that the observable changes in such an investigation will be shifts in the frequency distributions of the traits selected for study – the flies will not suddenly change into something else. As new combinations of traits accumulate, the changes in a species might be significant, but the fossil record tells us that dramatic changes in a species cannot be observed within a human lifetime.

The basic design of this exercise is to maintain six different population cages for several weeks. Each cage will begin with a defined population of flies, and the population will be monitored every generation to see how the distribution of traits changes. Ideally the exercise should be performed in several classes so that the results of replicate cages can be added to give a large sample size. As we shall see in chapter 15, sample size is a very important factor in studies of evolution.

An investigation following this design can be performed in many ways depending on what resources are available. Figure 14.3 illustrates a simple plexiglass cage designed for this exercise. Steps 1 to 3 refer specifically to this kind of cage. Other types of cages will also work, even vessels as simple as large bottles with foam plugs. Your instructor will show you what apparatus you will use and what methods of anesthetizing and handling flies you will employ. Modify the instructions as needed to suit your situation.

Each group will maintain a cage with one of the following populations:

Controls

These cages will start out with 20 flies of a single type – either completely wild or completely mutant. There will be no evolution in these cages. They serve to show how the population of each type of fly will grow in the absence of competition. There will be three controls: wild, mutant 1, and mutant 2.

Single mutant competition

These cages start with a population of 10 wild flies and 10 mutant flies. Therefore the initial allele frequency of the mutant will be 50%. You will see if it changes due to competition with

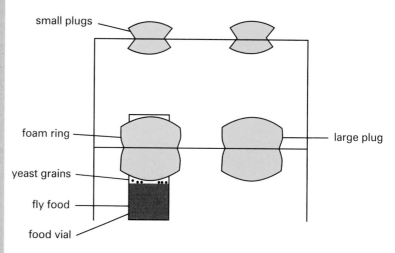

Figure 14.3 *A fruit fly population cage*

the wild flies. There will only be one mutant allele per cage, so there will be two of these cages, one for each mutant.

Double mutant competition

This cage will be started with 10 wild flies and 10 flies exhibiting *both* mutant traits. At the start each mutant allele frequency will be 50%. At first, the flies will either exhibit both mutations or none at all, but as the population evolves, you may find flies showing just a single mutation. Therefore you will need to count how many flies have each mutation, how many have both, and how many have neither. Check to be sure your subtotals equal the total number of flies for each count. This cage will allow us to investigate whether natural selection can promote new genetic combinations in a population.

Week One

1.   The class will be divided into six groups. Each group will maintain its own cage with one of the populations listed previously. The top of each cage is removable and has two 1-inch holes in it. The bottom of each cage has two 1.5-inch holes. Plug the small holes in the top with small foam plugs. Place a foam ring in one of the large holes in the bottom of the cage.

2.   Place 7.5 ml (1/2 tablespoon) of instant fly food into a glass vial and add 10 ml of water. In a few minutes the food should absorb the water and form a moist blue gel about 3 cm high. Sprinkle a few grains of yeast onto the gel (but don't cover it

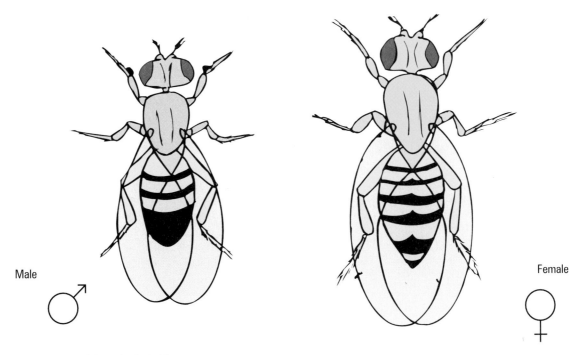

Male   Female

Figure 14.4 *Male and female fruit flies*

completely!), and insert the vial into the foam ring. (The fermentation products of the yeast are an integral part of the fruit fly's diet.) The vial should fit snugly with no gaps through which flies might escape.

3. Plug the other large hole with a large plug for now; in a few weeks you will add a second vial of fresh food in this opening. The cage should resemble figure 14.3.

4. After the cage is prepared, obtain some anesthetized flies from the appropriate culture stocks. Separate 20 females (as illustrated in figure 14.4) from the pool of anesthetized flies. No males will be added to the cultures, but it is safe to assume that most of the females have mated and are ready to lay eggs. This procedure eliminates the need to collect virgin flies, which is tedious and difficult for beginners.

5. Place the flies in the plastic portion of the cage. Do not put them directly into the food vial where they might suffocate in the soft food.

6. Put the top of the cage on and tape it down so it will not fall off. It is not necessary to seal the top completely with tape unless obvious gaps are visible. Be sure the cage is labeled with your name, the date, and what kind of flies your group is using. Place the cage where your instructor directs, taking care when you move it that the food vial does not fall out of the foam ring.

7. Construct a table to record the data for this experiment. There should be a column for the date of each entry, the number of wild phenotypes counted, the number of mutant phenotypes counted, and the calculated frequency of the mutant allele. There should be space in the table for at least eight entries (today's being the first). The class schedule will dictate exactly how many counts take place.

8. In the first entry, simply record how many mutant and wild flies you added to the cage. Because these flies were all taken from pure stocks, they are all homozygotes. Therefore the frequency of the mutant allele is the same as the percentage of mutant flies.

9. In the cages involving competition, many possible factors might conceivably exert pressure on differing levels of fitness — e.g., temperature, food, etc. List as many of these factors as you can. Whatever selection you may observe will be the result of how the different flies react to *all* of these factors. Different environmental conditions might cause selection to move in a different direction. Determining which factors are most important to the results you obtain would require numerous controls. How would this affect the experimental design? How does this exercise compare with exercise C in chapter 13?

Week Two

After the flies have had time to lay their eggs, remove the adult flies. Note if any of the flies were dead (which probably indicates they never recovered from being anesthetized). Make a corrected entry in your table if necessary to indicate how many live flies were initially placed in the cage.

Week Three

1.   When the second generation adults emerge (about two weeks after beginning) you will count the new numbers of each phenotype. Since no males were added to the cage when you set it up, the second generation will still be all homozygotes (the inseminations all occurred in pure culture). Therefore the frequency of the mutant allele will still be the same as the percentage of mutant flies.

2.   After you count the second-generation flies, remove them as you did their parents. Their eggs are in the food medium and represent the first generation in which the genes of the two strains are mixed.

Subsequent Weeks

1.   Continue the experiment through several generations. Count the phenotypes each generation, about every two weeks.

2.   After counting the third generation, remove the large plug from the bottom of the cage and replace it with a foam ring. Add a second food vial to the cage and replace the counted flies. Allow the population to grow from now on and let competition drive selection.

3.   Beginning with the third generation, the population will include heterozygotes. Use the Hardy-Weinberg equation from chapter 11 to estimate the allele frequencies based upon the observed frequency of the recessive phenotypes. Since this is not an ideal population and is subject to selection, the Hardy-Weinberg equation does not apply perfectly. Therefore you may occasionally see evidence that your calculated estimates are off (for example, one week you may see no recessive phenotypes and conclude that the allele has disappeared, only to see it reappear in the next generation). However, the estimates will be close enough to show the general trends over several generations.

Analysis

Monitoring this experiment is simple – you will just count the flies of different phenotypes every two weeks. *Understanding* what you are seeing will be more difficult. Eventually we will have to compile the data from all of the cages maintained by all of the classes doing this experiment in order to get a clear idea of what happens.

The fundamental question we are examining in this experiment is whether the cages are in Hardy-Weinberg equilibrium with respect to the meristic traits we have chosen to study. If they are, our hypothesis is that the ratio of wild to mutant alleles will remain at 50/50 over time. If they are not in equilibrium, we should expect to see a change in allele frequency. Because the Hardy-Weinberg theorem gives us a prediction of what to expect at equilibrium, we can use chi-square analysis to see if our final observed allele ratio is significantly different from 50/50.

If there is a significant change in allele frequency, we may conclude that our populations are not at equilibrium. We will discuss the forces that can keep a population from equilibrium in the next chapter.

A NOTE ABOUT RECORD KEEPING:
When working with live organisms, the unexpected often happens. Be sure to record any unplanned or unanticipated events in your notes, as these may be very important in interpreting your results later. (For example, if some of your mutants accidentally die, it will certainly alter the mutant allele frequency in future generations!)

IMPORTANT TERMS

Drosophila melanogaster	evolution	mutant	wild type

Laboratory Report Variation in Meristic Traits

FACTUAL CONTENT

Define the following terms. Brief definitions are given in the chapter, but you may also wish to consult your lecture notes and text for a better idea of how the terms are used. Pay close attention to those words that look similar but have different meanings – you will want to be able to tell them apart if they appear next to each other on a multiple-choice exam.

Drosophila melanogaster

evolution

mutant

wild type

EXERCISES - Record your data and answer the questions below.

A. Tabulating meristic variation in human fingerprints

Record the distribution of each pattern, by sex, for the whole class:

	female	male
arch		
radial loop		
ulnar loop		
whorl		

Are the patterns distributed differently among the sexes?

Construct a bar graph of the data.

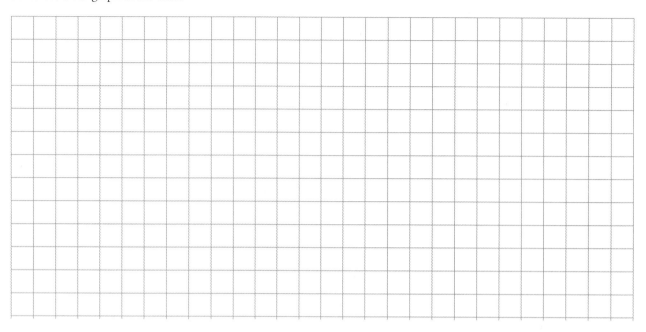

How does this bar graph compare to the histograms you constructed in chapter 13?

B. Studying shifts in meristic variation over time

Record what kind of flies you will be maintaining in your population cage:

Keep a record of your fly counts in the table.

date	number of wild flies	number of mutant flies	frequency of the mutant allele

List as many factors that might exert pressure on differing levels of fitness as you can:

Determining which factors are most important to the results you obtain would require numerous controls. How would this affect the experimental design? How does this exercise compare with exercise C in chapter 13?

When the experiment is complete, your instructor will total the counts from all of the replicate cages into a master data set. Use this to calculate the allele frequencies in the experimental cages for each week. Remember that in the early generations the flies are homozygotes and the allele frequencies are the same as the phenotype frequencies, but later you will need to use the Hardy-Weinberg equation to estimate the allele frequencies.

Week	Wild v Mutant 1	Wild v Mutant 2	Wild v Mutant 1 Mutant 2

Graph the mutant allele frequency for each trait versus the generation (see chapter 15 for model graphs).

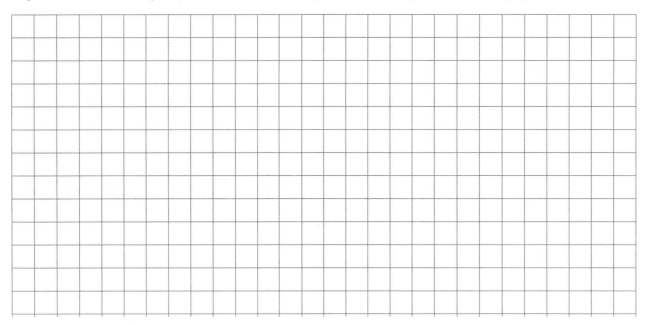

Do the mutant allele frequencies shift significantly from the starting values? How do the different mutants compare? Is there evidence of selection?

Use the master data set for the control cages to construct growth curves for the wild flies and each mutant strain in the absence of competition.

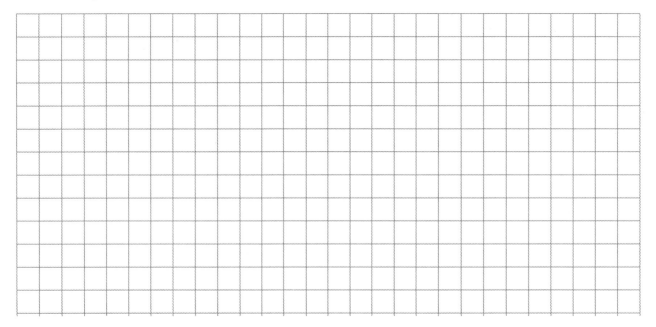

Do the control populations give you any ideas for why the experimental populations changed as they did?

Selection vs. Genetic Drift

15

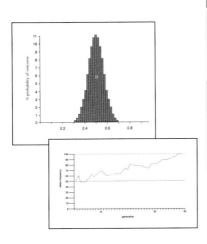

INTRODUCTION

The idea of natural selection follows logically from two premises:
1. The individuals of a population vary from each other.
2. Populations tend to produce more offspring than can possibly survive under normal circumstances.

Both of these premises are supported by empirical evidence. As we have seen in previous chapters, phenotypic variation is measurable and common. Naturalists have also long known that organisms reproduce prodigiously – a single tree may produce thousands of seeds, and an insect may lay hundreds of eggs. It stands to reason that if not all offspring can survive, only a sampling of them will.

One possibility is that the survivors will vary in the same way and in the same proportions as their parents – that is, that they will be a faithful sample of the parental generation. This situation would lead to no changes over time – it represents the Hardy-Weinberg equilibrium. Darwin reasoned that this would not always happen. Instead, those individuals whose variation gave them some advantage would be more likely to survive, slowly changing the distribution of variation.

Darwin's logic was good, but was he right? In chapter 1 we noted that logic can lead us to useful hypotheses, but we cannot consider anything as "scientifically factual" until we have a body of verifiable observations on the subject. In the case of natural selection, there are many such observations. The rise of antibiotic-resistant microbes is common evidence of natural selection. The fossil record, with its archive of transitions from primitive to derived forms, is another kind of evidence. But is natural selection the only force acting here?

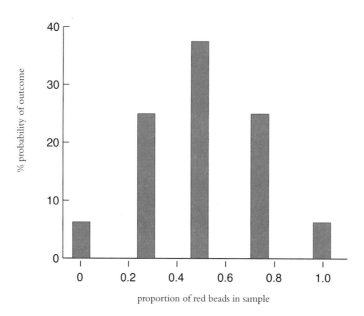

Figure 15.1 *Frequency of outcomes from sampling beads four at a time*

STATISTICS AND SAMPLING ERROR

Let us return to the notion that each generation is a *sample* of the variation present among its parents. The science of statistics and probability was devised specifically to deal with the question of how samples relate to the populations they are taken from. In the 1930s, the geneticist Sewell Wright reviewed the theory of natural selection in the light of probability theory and came to a startling conclusion: regardless of selective forces, all populations must also be subject to certain random tendencies. Because of this, one allele might occasionally replace another no matter what their relative effects on fitness. Wright called this random process **genetic drift**. The basis for the theory of genetic drift is the simple fact that samples are seldom *exact* representations of the population being sampled. This is the same phenomenon that guarantees that polls and surveys will show some margin of error no matter how carefully the pollsters choose their subjects.

The deviation of a sample from the parent population is called **sampling error**. The term *error* is a bit misleading – a sampling error is not a mistake, it is simply a normal deviation that can be expected in any random sample. Genetic drift is the result of sampling errors in the gene pool from generation to generation.

Sampling error is strongly influenced by the size of the sample taken, a situation best illustrated by a simple analogy. Imagine we have a large bowl of beads, half of which are red and half of which are white. If we pull one bead from the bowl at random, the probability that it will be red is 50 percent. By extension, if we pull several beads from the bowl at random, we would expect 50 percent of them to be red. Because the beads are removed at random, however, we will often find that the number of reds is a few more or a few less than half – that is, sampling error will occur.

When we first encountered genetics in chapter 11, we saw that through statistics we could actually predict the likelihood of a given deviation from an expected outcome. Some deviations are quite probable and do not make us question our expectations. Larger deviations are less likely, and if they are highly improbable, they imply that our expectations were flawed.

We can use probability theory to make some predictions about how much sampling error will occur in specific cases. In the case of four beads from a bowl, the proportion of red beads we observe in our sample must be one of five possibilities: 0% (no reds), 25% (one red), 50% (two reds), 75% (three reds), or 100% (four reds). A sample of two reds and two whites is the single most likely outcome, since it is most representative of the whole bowl of beads, but statistical calculations tell us that this result should only be expected in 37.5 percent of the samples. The remaining 62.5 percent of the time we would actually get something other than a 50/50 mix: 25 percent of the time there would be three reds, 25 percent of the time there would be one red, 6.25 percent of the time there would be four reds, and 6.25 percent of the time there would be no reds. This frequency distribution is illustrated in figure 15.1.

The situation changes markedly as the sample grows. Figure 15.2 shows the frequency distribution of the expected outcomes from pulling a sample of 50 beads from the bowl.

When 50 beads are sampled from the bowl, we still expect half of them (i.e., 25 beads) to be red. However, the shape of the distribution of possible outcomes is much narrower when 50 beads are removed than when only four are removed. This indicates that there is less variance in the outcomes. Thus, even though the 50 bead sample is exactly 50/50 only about 11 percent of the

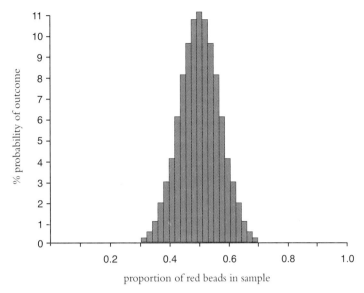

Figure 15.2 *Frequency of outcomes from sampling beads 50 at a time*

time, when it is not precisely 50/50 it is usually very close. This is reflected in a low standard deviation of 0.071. The standard deviation for the first distribution is over three times larger at 0.25. The magnitude of the standard deviations is proportional to the amount of sampling error that is likely to occur. We can see how much the sampling error drops with increasing sample size by noting that there is a 6.25 percent chance of getting all red beads when only four are removed, but when 50 beads are removed, the chance of them being all red plummets to 0.000 000 000 000 09 percent! Of course, all this math is just another way of stating the commonsense notion that a large sample is much more likely to be representative of the whole than a small sample.

GENETIC DRIFT AS SAMPLING ERROR

Every generation the genes in a population go through a sampling process analogous to our bowl of beads. Let's return to fruit flies for a parallel example, using the allele for scarlet eyes instead of red beads. Each fly can produce hundreds of sex cells, so even in a small vial of flies the actual population of potential offspring is quite large. However, not all of these gametes will be involved in making fertilized eggs, and not all of the eggs that are laid will hatch, and not all of the larvae will survive to reproduce. Therefore the alleles in the next generation will be just a sample of those present in the parents. If we start with a population in which the scarlet allele frequency is 50 percent, we would expect the frequency in the next generation to be the same. But just as with the beads, the probability of it remaining exactly 50 percent is small. If the second generation of flies is relatively large, the frequency will likely be close to 50 percent even if it is not exactly 50 percent. But if the second generation is relatively small, the deviation could easily be significant.

If this population is kept in a cage with a reasonable amount of food, the population will grow generation after generation. Even if there is no pressure that might select for or against scarlet eyes, sampling error will make the allele frequency change somewhat each generation. The second generation might have a scarlet frequency of 60 percent instead of the original 50 percent. The alleles in the gametes that will make up the third generation will also be 60 percent scarlet, but again, only a sample of this will actually become adult flies. Sampling error will again make it unlikely that the 60/40 ratio will be maintained exactly, and in the third generation we may see something different, say 48 percent scarlet. The allele frequency simply drifts without direction. Each generation the baseline shifts – that is, the next generation will be a sample of what the current gene pool is, not the original 50/50 mix of scarlet and wild. As the allele frequency drifts toward one extreme or another (either 100 percent scarlet or 100 percent wild) it becomes increasingly likely that a sample may contain only one type of allele. At that point the population will have lost its variability and drift ends. There is now only one possible allele, which is said to be **fixed** in the population. Figure 15.3 shows what might happen to the scarlet allele frequency in our hypothetical population. The population drifts back and forth until it has evolved from one of mixed eye color to one of uniform color. After generation 39 the wild allele has disappeared and the population contains only scarlet alleles. But this was not due to selection – under genetic drift the scarlet allele could just as easily have disappeared. What happened was that the population size was small enough that genetic drift prevented the maintenance of the Hardy-Weinberg equilibrium.

Given enough time, fixation of one allele over another would be an inevitable result of genetic drift (unless opposed by sufficiently strong selection or mutation pressure). The only case in which fixation would never occur is when population size is infinite.

157

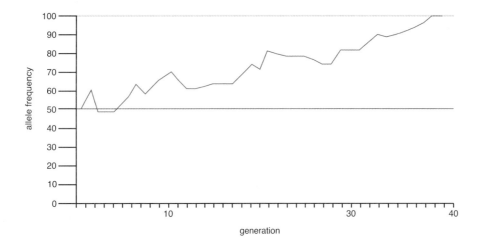

Figure 15.3 *Drifting frequency of scarlet allele in a hypothetical population*

Since all populations are considerably smaller than this, drift must be a force in evolution to some extent. Theory predicts that in large populations of thousands of individuals, drift is small, and the number of generations needed to fix an allele by drift may be more than the average number of generations that a species exists on earth. In small, isolated populations, especially those of 100 or fewer individuals, drift can be large, and fixation quite rapid. The number of generations needed for an allele to become fixed in a population by drift is thus proportional to population size.

SELECTION AND DRIFT

It is possible for a gene frequency to be simultaneously affected by both genetic drift and natural selection. The previous example was contrived, as it is known that under most conditions fruit flies with scarlet eyes do not survive as well as wild flies. A large population of mixed eye colors will become fixed on the wild allele at a fairly predictable rate that is proportional to the relative fitnesses of the two alleles. In small populations, drift will be more significant and the rate of fixation less predictable. In very small populations, drift can be large enough to occasionally overcome selection, even when the selective advantage of the allele is huge.

Biologists still debate the relative importance of selection and random factors like drift in the earth's evolutionary history. Many believe that, because drift is most important in small populations, overall its importance is slight and that most observable species characteristics are due to adaptive selection. Others argue that the vagaries of history are such that drift in populations at key times can and does influence evolution in significant ways. When scientists undertake long-term studies of wild populations, they try to collect data that may provide answers to such questions.

EXERCISES WITH GENETIC DRIFT

The following exercises are all simulations of drift as sampling error based on the bowl of beads analogy. They can be performed rather quickly. The principles these exercises illustrate can then be applied to your ongoing investigation of fruit fly populations begun in chapter 14.

A. Drift in a very small population

This simulation uses a sample size of four beads. If you think of each bead as an allele in a diploid organism, then it represents a case in which the population is limited to two individuals, but all that really matters here is how the frequency of bead colors changes over time.

1. Pull four beads from the jar marked "50/50" and count the number of beads of each color. Record the new frequency and replace these beads in the 50/50 jar.

2. For the next generation, go to the jar that represents the color frequency you drew from the 50/50 jar. That is, if you drew two beads of one color and two of the other, continue using the 50/50 jar, but if there were three beads of one color, then use the 75/25 jar for the next generation, etc. For each generation use the jar that reflects the ratio of bead colors drawn in the previous generation. Always return the beads to the jar they came from so that jar's population does not change.

3. Record the ratio of colored beads for every generation, and continue until all the beads in a sample are of one color (that is, the "allele" is fixed). Note how many generations it took to achieve fixation.

4. Repeat steps 1 to 3 for a second trial. Does the system fix on the same color? How does the number of generations to fixation compare?

5. Graph the results of both trials as a plot of the frequency of one of the bead colors at successive generations. Your graph should resemble figure 15.3. Plot both trials on a single axis.

6. Your instructor will tabulate the data from the entire class. What is the average number of generations to fixation? How many trials fixed on each color? Each *individual* trial results in four beads of one color and none of the other. What happens if you add all of the results together? Is the ratio of colors in all of the trials taken together close to 50/50?

B. Drift in a larger population

1. Follow the same procedure as in exercise A, but this time take samples of 10 beads. For 10 beads at a time there are more potential ratios, so be sure to use the set of jars specifically for this exercise. It has jars marked "50/50," "60/40, "70/30," etc. instead of the 25/75, 50/50, and 75/25 jars used for exercise A.

2. Do this exercise twice. Record the color of fixation and number of generations to fixation for each trial.

3. Graph the results of both trials as in part A.

4. Your instructor will again tabulate the data from the entire class. What is the average number of generations to fixation? How does this compare to the smaller population? How many trials fixed on each color? Again, what happens if you add all of the results together – as a whole, is the sum close to 50 percent of each color?

C. Drift and selection

1. Follow the procedure for exercise B, only use the bowls filled with different-sized beans instead of the colored beads. Do this experiment twice and graph the results as before.

2. Again, compare your results to those of the class. What happens if you add all of the results together this time? Is there consistent evidence of selection? Judging by the number of generations needed for fixation, is the "selection pressure" large or small? Would this influence the relative importance of selection versus drift?

D. Applying the principles

1. Consider an experiment following the design of the fly selection exercise in chapter 14. The investigation is being done in 10 different classes, and the results from all of the classes will be added together – that is, all of the cages involving competition between wild flies and flies with the first mutant will be treated as one giant population, etc. If each cage were experiencing drift but not selection, what kind of change in allele frequencies would you expect to see in the populations when added together? (How does this compare to exercise A, step 6?)

2. If each cage were experiencing consistent selection, what kind of change in allele frequencies would you expect to see in the populations when they are added together? (How does this compare to exercise C, step 2?)

3. Use these hypotheses to evaluate your final results in the fly investigation. Do you see evidence of selection or drift?

IMPORTANT TERMS

fixation   genetic drift   sampling error

Laboratory Report Selection vs. Genetic Drift

FACTUAL CONTENT

Define the following terms. Brief definitions are given in the chapter, but you may also wish to consult your lecture notes and text for a better idea of how the terms are used. Pay close attention to those words that look similar but have different meanings – you will want to be able to tell them apart if they appear next to each other on a multiple-choice exam.

fixation

genetic drift

sampling error

EXERCISES - Record your data and answer the questions below.

A. Drift in a very small population

Record the percentage of colored beads for every generation until all the beads in a sample are of one color:

	percentage of colored beads				percentage of colored beads	
generation	first trial	second trial		generation	first trial	second trial

Graph the frequency of colored beads versus the generation. Graph both trials on one axis.

Do both trials fix on the same color?

How does the number of generations to fixation compare?

Your instructor will tabulate the data from the entire class. What is the average number of generations to fixation?

How many trials fixed on each color?

What happens if you add all of the results together? Is the ratio of colors in all of the trials taken together close to 50/50?

B. Drift in a larger population

Record the percentage of colored beads for every generation until all the beads in a sample are of one color:

	percentage of colored beads				percentage of colored beads	
generation	first trial	second trial		generation	first trial	second trial

Do both trials fix on the same color?

How does the number of generations to fixation compare?

Graph the frequency of colored beads versus the generation. Graph both trials on one axis.

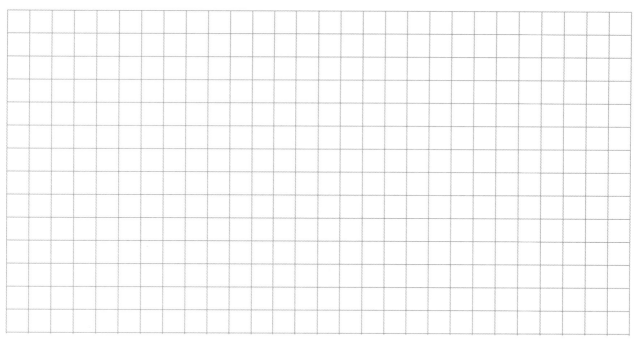

Your instructor will tabulate the data from the entire class. What is the average number of generations to fixation? How does this compare to the smaller population?

How many trials fixed on each color?

What happens if you add all of the results together? Is the ratio of colors in all of the trials taken together close to 50/50?

C. Drift and selection

Record the percentage of large beans for every generation until all the beans in a sample are of one size:

percentage of large beans

generation	first trial	second trial

Graph the frequency of large beans versus the generation. Graph both trials on one axis.

Do both trials fix on the same type of bean?

How does the number of generations to fixation compare?

Your instructor will tabulate the data from the entire class. What is the average number of generations to fixation? How many trials fixed on each type of bean?

What happens if you add all of the results together? Is the ratio of bean types in all of the trials taken together close to 50/50?

Is there consistent evidence of selection?

Judging by the number of generations needed for fixation, is the "selection pressure" large or small? Would this influence the relative importance of selection versus drift?

D. Applying the principles

Consider an experiment following the design of the fly selection exercise in chapter 14. The investigation is being done in 10 different classes and the results from all of the classes will be added together — that is, all of the cages involving competition between wild flies and flies with the first mutant will be treated as one giant population, etc. If each cage were experiencing drift but not selection, what kind of change in allele frequencies would you expect to see in the populations when added together? (How does this compare to exercise A?)

If each cage were experiencing consistent selection, what kind of change in allele frequencies would you expect to see in the populations when added together? (How does this compare to exercise C?)

Chemical Cycles and Energy Flow

16

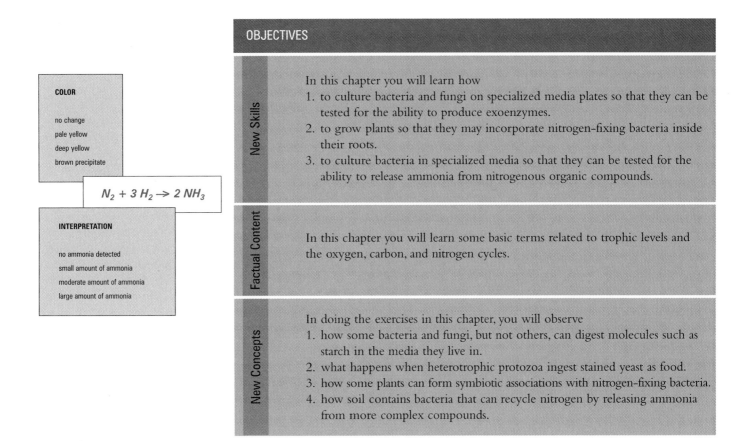

COLOR

no change

pale yellow

deep yellow

brown precipitate

$$N_2 + 3 H_2 \rightarrow 2 NH_3$$

INTERPRETATION

no ammonia detected

small amount of ammonia

moderate amount of ammonia

large amount of ammonia

OBJECTIVES

New Skills

In this chapter you will learn how
1. to culture bacteria and fungi on specialized media plates so that they can be tested for the ability to produce exoenzymes.
2. to grow plants so that they may incorporate nitrogen-fixing bacteria inside their roots.
3. to culture bacteria in specialized media so that they can be tested for the ability to release ammonia from nitrogenous organic compounds.

Factual Content

In this chapter you will learn some basic terms related to trophic levels and the oxygen, carbon, and nitrogen cycles.

New Concepts

In doing the exercises in this chapter, you will observe
1. how some bacteria and fungi, but not others, can digest molecules such as starch in the media they live in.
2. what happens when heterotrophic protozoa ingest stained yeast as food.
3. how some plants can form symbiotic associations with nitrogen-fixing bacteria.
4. how soil contains bacteria that can recycle nitrogen by releasing ammonia from more complex compounds.

INTRODUCTION

The way in which species interact with each other has a great deal to do with how physical resources are used by living organisms. Chemically, the earth is essentially a closed system, the amount of material leaving the earth or falling to it from space being negligible. It is therefore imperative that organisms recycle the chemical elements that they are made of. Otherwise, atoms would be forever locked inside the corpses of dead organisms, and new life would be impossible.

On the other hand, energy cannot be recycled. As energy takes its various forms and passes through different chemical compounds, a portion is wasted with each transformation until at last there is no usable energy left. Thus there is a downward cascade from an external source of energy (i.e., the sun) through all the levels of an ecosystem, with each organism in the system using what energy it can to live. As we shall see, this one-way flow of energy is intimately linked with the cyclical flow of chemical elements.

THE CARBON CYCLE

We took our first look at chemical cycles when we considered cellular respiration and photosynthesis. The **oxygen cycle** is fairly simple: Photosynthesizing organisms remove hydrogen atoms from water molecules for use in reducing carbon dioxide. Oxygen is a by-product of splitting water and is released into the atmosphere. Aerobic organisms use oxygen to oxidize the organic compounds made in photosynthesis and end up converting it back into water. The **carbon cycle** follows a similar pattern but is more complex because there is such a multitude of different organic intermediates. Nevertheless, the carbon dioxide that is generated by cellular respiration is the same carbon dioxide that is the raw material for photosynthesis.

Organisms that manufacture their own organic molecules from carbon dioxide are called **autotrophic**. (The root -*troph* is derived from the Greek *trepho*, meaning "to nourish" and is seen in many ecological terms.) Other creatures are **heterotrophic** – they obtain organic molecules in the food they consume. Strictly speaking, heterotrophy and autotrophy refer specifically to how a creature obtains its carbon, but often the terms are used to describe an organism's energy source as well. That is because most heterotrophs can only obtain energy by digesting organic molecules via glycolysis and cellular respiration, while most autotrophs obtain their energy from the sun. Some bacteria are exceptions to this rule (e.g., chemoautotrophs use carbon dioxide for carbon but obtain energy from inorganic substances like iron or sulfur instead of the sun), but they are rare. Photosynthetic autotrophs are important enough to have been given their own special name, **phototrophs**.

While the world of autotrophs is dominated by a single type, the phototrophs, three kinds of heterotrophs are common. They are distinguished by how they obtain food. **Holotrophs** eat other organisms and digest their whole tissues to release nutrients. Most animals are holotrophs. (*Carnivores* are holotrophs that eat other animals, *herbivores* eat plants, and *omnivores* will eat almost anything.) Other heterotrophs are **parasites**. They live inside, or extend absorptive structures into, other organisms and take nutrients from the host's body fluids.★ Parasites are found in every kingdom of life: bacteria, protists, fungi, plants, and animals can all be parasites. The third option for heterotrophs is **saprotrophy**. Saprotrophs live off of dead or decaying matter. Most bacteria and fungi are saprotrophs. These organisms are found throughout the environment and are responsible for spoilage and decomposition.

Saprotrophs have a big problem: They sit on top of their food (or

★ The precise meaning of *parasite* can vary with context. As used here it describes a kind of heterotrophy. When ecologists talk about how organisms interact with each other, they use *parasite* to refer to any organism that lives at the expense of another. This latter definition may include creatures that do not live inside other organisms.

in it) and must consume it without actually eating. The only way to do this is to digest the food while it is still outside their bodies and then absorb the nutrition that is released. This is accomplished by secreting **exoenzymes** into the food. Exoenzymes are special enzymes that exit an organism's body in order to work on substrates in the environment. Exoenzymes are critical tools in any chemical cycle.

Ecologists often talk of **trophic levels** in an ecosystem. In this case they are referring more to how energy flows through the system than how carbon is cycled. The first trophic level in any ecosystem is that of the **producers**. Producers are the autotrophs that bring energy into the biosphere. As noted, photosynthetic bacteria, algae, and plants are the most important producers on earth. The other trophic levels are various types of **consumers**. **Primary consumers** eat the producers, whereas **secondary** and **tertiary consumers** eat other consumers. The final trophic level, the **decomposers**, are the saprotrophs that digest the wastes and detritus at any level of the system. Thinking in terms of levels of consumption is important because of the principle that energy is always wasted as it flows through a system. For this reason each trophic level must be smaller in aggregate mass than the one beneath it – that is, carnivores such as lions are rarer than the herbivores they hunt, and acres of vegetation are necessary to support herds of grazers.

THE NITROGEN CYCLE

The element nitrogen is a component of both amino acids and nucleotides and is therefore critically important to life. Although nitrogen gas makes up nearly 80 percent of our atmosphere, only a few kinds of bacteria can use gaseous nitrogen directly. These bacteria are able to reduce nitrogen gas by adding hydrogen atoms to make **ammonia**:

$$N_2 + 3\,H_2 \rightarrow 2\,NH_3$$

This reaction is actually quite complex, requiring several enzymes and ATP as an energy source. The entire process is called **nitrogen fixation**. Some cyanobacteria can fix nitrogen, and since they are also autotrophs, they are arguably the most self-sufficient kind of life on earth.

The most powerful nitrogen-fixers are not cyanobacteria, but heterotrophic bacteria that live in the soil. Some of these can live symbiotically inside plant roots where they can "trade" their ammonia for some of the plant's carbohydrates. Only certain species of plants can accept these bacterial partners; they form **nodules** on their roots to house the nitrogen-fixing bacteria. Many legumes, including common clover, form these nodules.

Just as all life depends on autotrophs to make organic compounds out of carbon dioxide, the world also depends on nitrogen-fixing bacteria to make ammonia. The inability of plants to use nitrogen on their own means that it is often a limiting factor for plant growth. This is why farmers spread tons of nitrogen fertilizer on crop land every year.

Of course, plants have not always depended on fertilizers for their nitrogen. Nature is very good at recycling the nitrogen that is already in ecosystems so that it can be reused. When heterotrophs eat plants and metabolize the proteins in them, they usually release some of the nitrogen as ammonia, which is a waste product in animals (urine contains a great deal of ammonia and ammonia-like compounds). Furthermore, when organisms die, their tissues are attacked by various bacteria, some of which metabolize the proteins present. These bacteria also produce ammonia, and this decay process is called **ammonification.** Nitrogen fixation, the accumulation of wastes, and ammonification all mean that nitrogen is constantly entering the soil in the form of ammonia.

Some types of bacteria in the soil can oxidize ammonia for energy just as other organisms oxidize carbohydrates. When oxidized, ammonia turns into nitrate. In fact, many plants absorb most of their nitrogen as nitrate instead of ammonia. The process of converting ammonia into nitrate is called **nitrification**. Together, nitrogen fixation, ammonification, and nitrification complement each other in the **nitrogen cycle**.

EXERCISES WITH THE CARBON AND NITROGEN CYCLES

You already encountered some aspects of the carbon cycle in chapters 8 and 9 on cellular respiration and photosynthesis. Exercises A and B focus on how saprotrophs digest organic molecules and how holotrophs ingest and break down other organisms. Exercises C and D highlight two aspects of the nitrogen cycle. Exercise C demonstrates how some plants can form nodules of symbiotic bacteria to fix nitrogen, while exercise D involves other kinds of soil bacteria that can recycle nitrogenous compounds into ammonia.

A. Starch exoenzymes from saprotrophs

As you may recall, the polysaccharide starch turns a deep blue color when stained with a solution of iodine/potassium iodide, but mono- and disaccharides do not. In the following test, you will grow bacteria and fungi on agar plates containing starch. Those organisms that can make an exoenzyme to digest starch will produce an area on the plate where the starch has turned to sugar. These areas will remain clear when the plate is stained with iodine solution.

This test is specific for exoenzymes that digest starch. Other media would be required to test for exoenzymes that digest proteins or other compounds.

Materials

starch agar plates	nutrient agar plates containing starch
bacterial cultures	pure cultures of *Bacillus megaterium*, and *Escherichia coli*
fungal culture	*Penicillium* sp. grown on rose bengal or Saboraud agar until blue conidia form
IKI solution	

PERIOD ONE

1. Draw a line on the back of a starch agar plate so that it is divided into two equal areas. Label one side *E. coli* and the other *B. meg.*

2. Flame a loop and aseptically transfer a loopful of *E. coli* to the middle of the appropriate region of the plate. Do not spread the culture around – leave it in a small area.

3. Repeat step 2 for *B. megaterium*.

4. Now take a second starch agar plate and mark the back *Penicillium*.

5. Wipe a sterile swab through the blue-green spores of the *Penicillium* and transfer the spores to the plate. Best results are obtained if you inoculate the plate with a circle of spores in the center, so the fungus will grow in a ring.

6. Place the plates where your instructor directs.

PERIOD TWO

1. Flood each plate with IKI solution.

2. Compare the results for *E. coli* and *B. megaterium*. Which one has digested the starch? Can the *Penicillium* digest starch?

B. Predation by heterotrophic protozoa

Protozoa are unicellular creatures that are usually heterotrophic. Most of them swim around in water hunting and eating smaller cells, acting like tiny animals (the term *protozoa* comes from Greek words meaning "primitive animals," although few protozoa seem to have any genetic relationship to true animals). Yeast cells are small enough to be eaten by many protozoa. Yeast cells are colorless but can be stained with the dye Congo red.

Materials

protozoa cultures	pure cultures of *Paramecium* or *Didinium* in hay infusion
yeast stock	boiled baker's yeast stained with Congo red

1.   Mix some yeast with paramecium culture and make a wet mount slide of the preparation. (This works best if some of the debris from the hay infusion is put on the slide so that the paramecia will stick to it instead of swimming around – they can be very fast!)

2.   Examine the slide under the microscope and look for paramecia that are ingesting yeasts in their scoop-shaped "mouths." Note if the yeasts change color after they are ingested. Why might this happen?

C. Root nodules and nitrogen fixation

Materials

Rhizobium culture/ clover seeds	fine soil containing *Rhizobium* bacteria and clover seeds
seeds	clover, peas, and corn
crystal violet stain	

Rhizobium is a nitrogen-fixing bacterium that can form root nodules in some plants but not others. It takes weeks for the nodules to develop, so these plants will have to be allowed to grow before any results can be seen.

PERIOD ONE

1.   Fill four plastic pots with potting soil. Water all pots thoroughly.

2.   In two of the pots spread plain clover seed, and in the other two pots spread clover seed mixed with *Rhizobium* culture. Take care as you work so that you do not contaminate the plain seed with *Rhizobium*.

3.   Add two to three pea seeds to one of the pots with *Rhizobium* and one without. Add two to three corn seeds to the other pots. You should now have pots containing the following:
   a.   clover plus peas
   b.   clover plus *Rhizobium* plus peas
   c.   clover plus corn
   d.   clover plus *Rhizobium* plus corn
What is the variable in this experiment? What are the controls?

4.   Place a marker in each pot to show what is planted in it and whether it contains *Rhizobium*.

5.   Cover the seeds with about a centimeter of soil and moisten the top thoroughly.

6.   Place the pots where your instructor directs. They should be watered regularly, but take care not to water so much that the *Rhizobium* cultures might be washed out of the pot. A plant mister works well for this.

PERIOD TWO   (five to six weeks after planting)

1.   Remove each plant from the soil and carefully rinse the roots with water to remove any dirt.

2.   Compare the roots. Which plants have nodules? Which are unable to make nodules?

D. Nitrogen cycling soil bacteria

Peptone broth is a bacterial culture medium that is rich in nitrogen-containing organic compounds. If a tube of sterile peptone broth is inoculated with ammonifying bacteria, those bacteria will be able to break down the nitrogenous compounds for food. The process releases ammonia, just as when these same bacteria cause materials in soil to decay. The ammonia can then be detected chemically with Nessler's reagent.

Materials

peptone broth	tubes of sterile bacterial medium
soil samples	samples of dirt or mud suspended in sterile water
Nessler's reagent	dilute HgI_2 in 20% KOH
ammonia standard	1% ammonium chloride

PERIOD ONE

1.   Obtain a sterile tube of peptone medium and a soil sample from your instructor. These samples were made by suspending soil in sterile water so that the water would pick up some of the bacteria present in the soil.

2.   Use a sterile pipet to take up 1 ml of the soil sample. Never mouth-pipet a sample of bacteria – use a pipettor. Transfer the bacteria to your peptone tube.

3.   Place the tube in the indicated place to incubate. Your instructor will also set aside a sterile tube of peptone broth to serve as a negative control.

PERIOD TWO

1.   Place one drop of Nessler's reagent into a depression of a spot plate. Add one drop of ammonia standard and note the color. The darker the color, the more ammonia is present:

COLOR	INTERPRETATION
no change	no ammonia detected
pale yellow	small amount of ammonia
deep yellow	moderate amount of ammonia
brown precipitate	large amount of ammonia

2.   Repeat step 1 with a drop from the control tube of sterile medium. Does it contain ammonia?

3.   Repeat step 1 with a drop from your bacterial culture. How does this differ from the sterile control? What effect did the bacteria have on the ammonia content of the medium? How does this relate to the nitrogen cycle?

IMPORTANT TERMS

ammonia	exoenzyme	nitrogen fixation	protozoa
ammonification	heterotrophic	oxygen cycle	root nodules
autotrophic	holotroph	parasite	saprotroph
carbon cycle	nitrate	phototroph	secondary consumer
consumer	nitrification	primary consumer	tertiary consumer
decomposer	nitrogen cycle	producer	trophic level

NAME

Laboratory Report Chemical Cycles and Energy Flow

FACTUAL CONTENT

Define the following terms. Brief definitions are given in the chapter, but you may also wish to consult your lecture notes and text for a better idea of how the terms are used. Pay close attention to those words that look similar but have different meanings – you will want to be able to tell them apart if they appear next to each other on a multiple-choice exam.

ammonia

ammonification

autotrophic

carbon cycle

consumer

decomposer

exoenzyme

heterotrophic

holotroph

nitrate

nitrification

nitrogen cycle

nitrogen fixation

oxygen cycle

parasite

phototroph

primary consumer

producer

protozoa

root nodules

saprotroph

secondary consumer

tertiary consumer

trophic level

EXERCISES - Record your data and answer the questions below.

A. Starch exoenzymes from saprotrophs

Sketch or describe the appearance of the plates after they have been treated with IKI solution:

Compare the results for *E. coli* and *B. megaterium*. Which one has digested the starch?

Can the *Penicillium* digest starch?

B. Predation by heterotrophic protozoa

Sketch or describe the paramecia that are ingesting yeasts:

Note if the yeasts change color after they are ingested. Why might this happen?

C. Root nodules and nitrogen fixation

What is the variable in this experiment?

What are the controls?

After the plants have matured, examine their roots for nodules and note which plants have nodules:

A)   clover plus peas

B)   clover plus *Rhizobium* plus peas

C)   clover plus corn

D)   clover plus *Rhizobium* plus corn

Which plants are unable to make nodules?

D. Nitrogen cycling soil bacteria

Record what type of soil you are testing in this exercise:

Record the results you obtained with Nessler's reagent on each sample:

sample	test results
ammonia control	
sterile peptone broth	
peptone broth + soil bacteria	

How does the soil bacterial culture differ from the sterile control?

What effect did the bacteria have on the ammonia content of the medium? How does this relate to the nitrogen cycle?

Bacterial Diversity and the Origin of Eukaryotic Cells

17

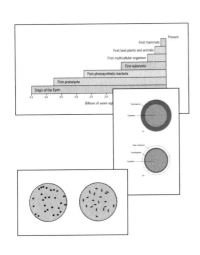

INTRODUCTION

The oldest rocks ever discovered are nearly four billion years old. Fossil impressions of cells are almost as old (see figure 17.1). As one might expect, these early cells were small and simple prokaryotes. In fact, prokaryotes have been the only kind of life for most of the earth's history. For millions of centuries they reproduced and diversified, developing into a variety of different kinds of bacteria. Some, such as oxygen-generating cyanobacteria, significantly altered the environment for future kinds of life. The first eukaryotes did not appear until after the bacteria had built up a two-billion-year head start and had already made their mark upon the planet.

All familiar forms of life – plants, animals, and fungi – are eukaryotic. They are relative newcomers to the earth. Eukaryotes remain vastly outnumbered by prokaryotes, and they have not adapted to as wide a range of environments as prokaryotes. On the other hand, eukaryotes have developed a tremendous diversity of forms and sizes in a relatively brief time. The story of how eukaryotic cells came to be and how they evolved into so many different kinds of organisms is one of the strangest discoveries of modern biology.

EARLY IDEAS ABOUT BACTERIA

Despite their antiquity, bacteria have only been known to humans for the last two or three centuries. No one knew what they were when they were first observed with microscopes – for some time they were simply thought of as "primitive plants." In the early days of biology, any creature that did not move, especially if its cells were surrounded by cell walls, was considered a plant.

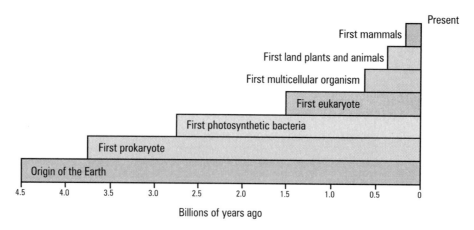

Figure 17.1 *The fossil history of life*

As people learned more, their ideas became more refined, and by the 1970s, the notion of dividing all life into plant and animal kingdoms gave way to a concept of five kingdoms of life. Bacteria, because of their prokaryotic cell structure, were given their own kingdom called **Monera**. The eukaryotes were divided into four other kingdoms because they were so diverse: the multicellular holotrophs became kingdom **Animalia**, the terrestrial autotrophs were designated kingdom **Plantae**, a complex line of saprotrophs became kingdom **Fungi**, and everything else (mostly unicellular) constituted the kingdom **Protista**. This five-kingdom classification is still followed by most biologists because it is convenient, but recent discoveries show that it does not really reflect the true scope of prokaryotic diversity nor how prokaryotes are actually related to eukaryotes.

ARCHAEA

Because bacteria are so similar to each other physically, it is very difficult for bacteriologists to determine who is related to whom and how they may have evolved. Early studies based on cell chemistry and physiological processes like fermentation could support only crude theories. Bacteriology began to change in the late 1970s when molecular biologists developed methods of sequencing DNA, opening a window directly into prokaryotic genetics. Using these techniques, Carl Woese, a molecular biologist at the University of Illinois, made an astounding discovery — there were actually two completely different kinds of prokaryotes.

Woese found that most familiar bacteria were similar enough to fall into a single group. He called this group the **eubacteria**, or "true bacteria." A second group of prokaryotes seemed only distantly related to the eubacteria. Their DNA was different, their cell walls were made of different materials, even the lipids in their cell membranes were odd. They also tended to live in strange environments: water that is extremely salty, boiling mineral springs, and other bizarre places. Many of these environments have little or no oxygen, so Woese guessed that the prokaryotes that lived there must be very similar to primitive life. Therefore

he named the group **archaebacteria**. It has now become common to refer to these creatures as **archaea**, in order to emphasize how different they are from eubacteria.

Originally Woese proposed making each of these groups its own kingdom. The eubacteria would include many familiar subgroups: the cyanobacteria, the nitrogen-cycling bacteria of the soil, the purple and green sulfur bacteria we saw in chapter 9, and many other types of bacteria. The archaea are divided into three main types: **methanogens**, anaerobes that release methane gas when they ferment; **halophiles**, which live in extremely salty water (*hals* is Greek for salt, *philos* Greek for loving, hence "salt-loving"); and **thermophiles**, which live at temperatures close to (or sometimes even exceeding) that of boiling water, conditions found in hot springs and underwater volcanoes. All evidence suggests that the eubacteria diverged from the archaea quite early in the earth's history, almost certainly before oxygen was plentiful. That is, these kingdoms have been growing apart since before eukaryotes even existed. They have very little in common today.

THE ENDOSYMBIOSIS THEORY

Woese made another strange discovery, one that helped to settle an evolutionary question first raised in the early 1900s. As you saw in chapter 3, two of the organelles found inside eukaryotic cells (mitochondria and chloroplasts) seem very similar to individual bacteria. This led some biologists to speculate that eukaryotes arose when some bacteria started living inside larger bacteria, an idea known as the **endosymbiosis theory**. The endosymbiosis theory was controversial; it was not clear whether it offered an explanation for the origin of the nucleus or other eukaryotic structures (most biologists who favored the theory limited it to mitochondria and chloroplasts) and therefore seemed to offer little to those seeking an explanation for the origin of eukaryotes. Nevertheless, the spirited advocacy of a few scientists made the theory popular among microbiologists during the 1970s. Unfortunately, it was almost impossible to test experimentally — that is, until the work of Woese.

Woese and his colleagues decided to see if mitochondria and chloroplasts were genetically related to either eubacteria or archaea. They could do this because both mitochondria and chloroplasts contain their own circular chromosomes, just like bacteria. They also contain tiny structures called ribosomes. Both contain nucleic acids that can be sequenced and compared for genetic relationships.

They found that mitochondria and chloroplasts were related to different groups of eubacteria. What was shocking was that the DNA and ribosomes of the rest of the cell were related to archaea! That is, part of a eukaryotic cell evolved from one type of prokaryote, and the other part evolved from the other kind of prokaryote. This essentially proved the endosymbiosis theory, for the only way this could happen was for the different parts to have come together after millions of years of being separate. The relationships between eubacteria, archaea, and eukaryotes are still being actively investigated, but the story at this point appears to be that the first eukaryotes arose from some kind of thermophilic archaea. The exact origin of the nucleus is obscure, but it is known that some thermophiles use specialized proteins to protect their DNA from heat, and similar proteins are also the basis for the scaffolding that supports eukaryotic chromosomes. At some point, perhaps driven by the increasing oxygen levels created by photosynthesizing cyanobacteria, these primitive eukaryotes picked up eubacterial symbionts that were able to perform very efficient aerobic respiration. Over time, these symbionts became mitochondria. Some time later, certain types of eukaryotes acquired photosynthetic cyanobacteria as chloroplasts.

The whole notion of "kingdoms of life" is now in flux. Microbiologists have proposed that life be divided into three *domains* of eubacteria, archaea, and eukaryotes. Each of these domains would then be subdivided into kingdoms – for example, Domain Eukaryota would contain the kingdoms Protista, Fungi, Plantae, and Animalia. On the other hand, eukaryotes are not really a separate domain, but a mixture of the other two. This, together with many complex details in the molecular data, have prevented biologists from reaching a consensus on a new classification system. As a result, the old five-kingdom scheme is still used alongside the new domain system.

EUBACTERIAL DIVERSITY

Although eubacteria are all more related to each other than they are to archaea or eukaryotes, they are still a large and diverse group of organisms. As we have seen, they can be phototrophs, chemoautotrophs, saprotrophs, or parasites. They can be aerobic or anaerobic. Those that ferment have many different methods of fermentation. Eubacteria also exhibit many structural and chemical differences too complex to describe in a basic biology course. Within their domain or kingdom are many subgroups that need to be identified and sorted out.

Yet this diversity is hidden by structural simplicity. With only a microscope it is often difficult to tell different species apart. In fact, most eubacteria are limited to one of just three possible shapes: rods (called **bacilli**), spheres (**cocci**), or spirals (**spirilla** and **spirochetes**). These basic shapes are illustrated in figure 17.2.

Bacteriologists have devised many kinds of tests for identifying bacteria in more detail. One of the most common tests they use is the **Gram stain**, named after its inventor, Hans Christian Gram. In the Gram stain, bacteria are first colored purple with a stain called crystal violet. This stain is then set with an iodine wash before being rinsed away with alcohol. **Gram-positive** bacteria will retain the crystal violet/iodine complex, but **gram-negative** bacteria will not. After the crystal violet/iodine stain, the sample is then stained red with safranin. Those cells already stained purple remain so; those that did not retain the purple dye will now be stained red. Thus the results are indicated by the final color of the bacteria.

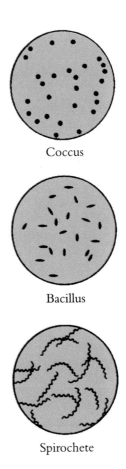

Coccus

Bacillus

Spirochete

Figure 17.2 Shapes of eubacteria

The purple and red colors of the Gram stain actually tell us something about the cell walls of the bacteria being tested. Gram-positive cell walls are thick but relatively simple. They are made of many units of a substance called **peptidoglycan**, stacked like bricks and "cemented" together into a sturdy, unified structure. Gram-negative walls only have a thin layer of peptidoglycan. But surrounding that is a second layer: an oily membrane that reinforces the peptidoglycan (see figure 17.3). These thinner walls do not retain the chemical complex formed by crystal violet and iodine like thick gram-positive cell walls do. This is why the alcohol wash in the Gram stain procedure decolors the gram-negative cells and allows them to be restained with safranin.

Cell wall composition is a useful characteristic in studying organisms. Only eubacteria have walls made of peptidoglycan; plant cell walls are made of cellulose, fungal cell walls are made of chitin, and archaea cell walls are made of a variety of substances. Of course, even within the eubacteria, there are many types of gram-positive and gram-negative bacteria. The gram-negative group includes many bacteria that affect humans. In fact, one reason that the Gram stain is used so much is because it turns out that gram-positive bacteria are usually more sensitive to penicillin and similar antibiotics than gram-negative bacteria are. While this generalization is not true in every case, it is accurate enough to make the Gram stain clinically important.

EUKARYOTIC DIVERSITY

The most ancient fossils of eukaryotic cells yet discovered are only about 1.5 billion years old – considerably younger than the oldest prokaryotes. Yet in these billion and a half years eukaryotes have evolved into forms that are structurally much more diverse than prokaryotes have achieved in nearly four billion years.

Eukaryotes generate phenotypic variation much faster than prokaryotes. This is because eukaryotic chromosomes are very different from prokaryotic chromosomes. As we saw in chapters 10, 11, and 12, eukaryotic chromosomes are long ribbons of DNA containing many "silent" regions; they pair up so that the alleles they carry can mix in various combinations, and crossing over can create even more genetic combinations. Furthermore, in each generation, the processes of meiosis and sexual reproduction produce still more new combinations. In contrast, prokaryotes generally have a single, small, circular chromosome. Although there are situations in which they can exchange bits of DNA, they have no chromosome pairs, no crossing over, no meiosis, no heterozygosity.

When eukaryotes appeared, they brought with them a new pace of evolution. It is impossible to look at examples of all of the different types of eukaryotes, so in this chapter we will focus on three unicellular types from the protist kingdom. In subsequent chapters we will look at other eukaryotes that shed light on evolutionary history. While it is important to remember that all of the specimens we look at are modern species, we can use them as models of what ancient creatures were like.

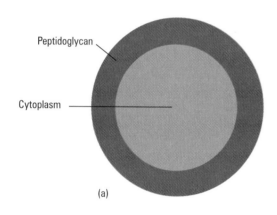

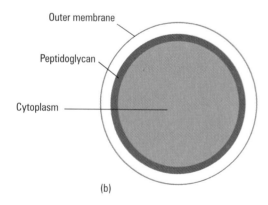

Figure 17.3 *Gram-positive (a) and gram-negative (b) cell walls*

EXERCISES WITH BACTERIA AND UNICELLULAR EUKARYOTES

A. Survey of archaea

The best known archaea live in very harsh environments. Because they require such difficult conditions for growth, you will not be asked to culture any. However, if available, some archaebacteria will be on display for you to examine. Describe what you see in your notes—do you see signs that they are living in an extreme habitat (temperature, salt crystals, etc.)?

B. Survey of eubacterial shapes

1. Obtain a prepared slide of mixed bacterial types. This slide has bacteria representing the three shapes of bacteria: bacilli, cocci, and spirilla. Under low power, find a portion of the slide that contains a diverse collection of cell types.

2. Switch to high power. Try to center a group that shows all three shapes. Next, raise the nosepiece so that the objectives are well clear of the slide. Switch to the oil immersion lens. Place a small drop of oil on the center of the slide. Now, watching from the side of the microscope, carefully lower the objective so that it makes contact with the oil. You should be able to focus the image with the fine adjustment knob. Adjust the diaphragm to sharpen the contrast.

3. Sketch each of the shapes in your notes. Do these shapes suggest much diversity? Do the various shapes come in different sizes?

C. Gram staining eubacteria

For this exercise you will make three different slides. *Bacillus megaterium* is gram-positive and *E. coli* is gram-negative – they will serve as controls. After you have made slides of those two, make a slide of *Enterococcus faecalis* and determine whether it is gram-positive or negative.

Materials

bacterial cultures	pure cultures of *Bacillus megaterium*, *Enterococcus faecalis*, and *Escherichia coli*

Gram reagents	solutions of crystal violet, Gram's iodine, safranin, and 95% ethanol

1. On the bottom of a clean slide, draw a circle about the size of a dime. This will be your guide.

2. Place a drop of water on the top of the slide over the circle.

3. Flame an inoculating loop to red-hot and use it to pick up a loopful of *B. megaterium*. Spread the sample around in the water, but try not to spread the area much larger than the circle. Flame your loop to sterilize it when you are done.

4. Let the slide air dry. Blowing or heating the slide at this stage will distort the specimen. If you spread your suspension thinly enough, this will not take long.

5. When the slide is completely dry, heat-fix the bacteria to the slide by passing it through a flame two or three times.

6. Hold the slide with a clothespin while you stain it. Cover the bacteria with crystal violet for 30 seconds. This stains all of the bacteria purple.

7. Hold the slide at an angle over a sink or staining tray and rinse excess stain from the slide with a gentle stream of distilled water from a wash bottle.

8. Cover the sample with Gram's iodine for 30 seconds. This fixes the violet stain.

9. Wash off excess iodine with alcohol. This also removes the violet stain from the gram-negative bacteria. Rinse the slide with water.

10. Cover the sample with safranin for 30 seconds. This re-stains the gram-negative bacteria red. Rinse with water, gently blot the slide with a paper towel, and let it air dry.

11. A heat-fixed slide of bacteria does not need a coverslip when it is dry. Oil can be placed directly on the sample.

12. Examine under the microscope. Record the shape and gram reaction of the sample.

13. Repeat the procedure for *E. coli* and *E. faecalis*. Is *E. faecalis* gram-positive or gram-negative?

D. Amoeboid motion in eukaryotes

All eukaryotic cells contain an internal framework of protein fibers called a **cytoskeleton**. This term is a bit misleading, however, for unlike the bony skeleton of a vertebrate, a cytoskeleton is not rigid but is capable of movement. Indeed, you have already seen evidence of this, for it is elements of the cytoskeleton that move chromosomes during mitosis and meiosis. In animals, muscle cells are modified so that they can use their cytoskeletons to contract to a fraction of their resting length, enabling the animal to move.

Amoebas are single-celled protists with such exquisite control over their cytoskeletons that they can change their shapes in order to crawl about and catch prey. Although the cytoskeleton is invisible under a light microscope, its actions can be seen in the movements of living amoebas.

1. Examine a culture jar of amoebas under a dissecting microscope and look for an individual amoeba. In water they look like tiny white stars. Draw it up with a pipet and make a wet mount slide.

2. Observe their structure under the microscope. Amoebas are colorless but can be stained with non-toxic dyes such as nigrosin, eosin, or cresyl blue to bring out more details.

3. Observe how the shape of the cell changes. Can you see the creature engulfing smaller cells as food?

E. Eukaryotic flagella

Many eukaryotic cells have whiplike extensions called **flagella**. (Some bacteria have flagella too, but prokaryotic flagella are structurally quite different from eukaryotic flagella.) Flagella generally function as propellers to push or pull the cell through liquid media – for example, many types of sperm cells swim to eggs using flagella. Several kinds of protists have flagella.

1. Examine a prepared slide of a trypanosome. These are parasites that live in blood.

2. Locate some long trypanosome cells amidst the round blood cells. Can you see the flagella?

F. Eukaryotic sex and reproduction: Ciliate conjugation

Most kinds of eukaryotes have some type of sexual reproduction (amoebas are a rare exception). The origin of sexual reproduction is hotly debated, but there is considerable evidence that suggests it arose from mechanisms required for the maintenance of eukaryotic chromosomes: Eukaryotic chromosomes literally fray at the ends and must periodically be repaired, a process that frequently happens during meiosis and subsequent sexual reproduction. This is unnecessary in prokaryotes, because their circular chromosomes have no ends to fray. Whatever its origins, sex is very important in eukaryotes, even unicellular eukaryotes.

In chapter 16 we looked at paramecia as examples of heterotrophs. Paramecia belong to a complex group of protozoa called *ciliates*. One characteristic of ciliates is that they have two types of nuclei: the **micronucleus** and the **macronucleus**. The macronucleus controls the cell, but the micronuclei can undergo meiosis and participate in sexual **conjugation**. When ciliates of opposite mating types (they do not have obvious sexes) conjugate, their macronuclei dissolve. Each gives the other cell a haploid micronucleus to fuse with one of its own haploid micronuclei. The resulting diploid micronucleus then gives rise to a genetically new macronucleus. After conjugation, the two cells separate. Thus, in ciliates, sex is separate from reproduction – two cells enter the process and two remain at the end. Conjugation allows the cells to alter their own genetic identity. Ciliates actually reproduce by cell division. Usually conjugation is followed by cell division.

1. Examine a prepared slide of paramecium division. Pay careful attention to the orientation of the dividing cells – do they divide lengthwise or end-to-end?

2. Now look at a slide of paramecium conjugation. How does this compare to their method of division? How does the orientation of the "mouths" facilitate the exchange of micronuclei?

IMPORTANT TERMS

archaea	endosymbiosis theory	halophile	Protista
archaebacteria	eubacteria	macronucleus	spirillum
Animalia	flagellum	methanogen	spirochete
bacillus	Fungi	micronucleus	thermophile
coccus	gram-negative	Monera	
conjugation	gram-positive	peptidoglycan	
cytoskeleton	Gram stain	Plantae	

17

Laboratory Report   Bacterial Diversity and the Origin of Eukaryotic Cells

FACTUAL CONTENT

Define the following terms. Brief definitions are given in the chapter, but you may also wish to consult your lecture notes and text for a better idea of how the terms are used. Pay close attention to those words that look similar but have different meanings – you will want to be able to tell them apart if they appear next to each other on a multiple-choice exam.

archaea

archaebacteria

Animalia

bacillus

coccus

conjugation

cytoskeleton

endosymbiosis theory

eubacteria

flagellum

Fungi

gram-negative

gram-positive

Gram stain

halophile

macronucleus

methanogen

micronucleus

Monera

peptidoglycan

Plantae

Protista

spirillum

spirochete

thermophile

EXERCISES - Record your data and answer the questions below.

A. Survey of archaea

Sketch or describe the archaebacteria on display:

Do you see signs that they are living in an extreme habitat?

B. Survey of eubacterial shapes

Sketch each of the bacterial shapes:

Do these shapes suggest much diversity?

Do the various shapes come in different sizes?

C. Gram staining eubacteria

Record the shape, color, and gram reaction of each sample:

	shape	color	gram reaction
Bacillus megaterium			gram-positive
Escherichia coli			gram-negative
Enterococcus faecalis			

D. Amoeboid motion in eukaryotes

Sketch or describe the amoebas under magnification:

Can you see the creature engulfing smaller cells as food?

E. Eukaryotic flagella

Sketch or describe the trypanosomes under magnification:

Can you see the flagella?

F. Eukaryotic sex and reproduction: ciliate conjugation

Sketch or describe paramecium division:

Do they divide lengthwise or end-to-end?

Sketch or describe paramecium conjugation:

How does this compare to their method of division?

How does the orientation of the "mouths" facilitate the exchange of micronuclei?

Algae and the Origin of Plants

18

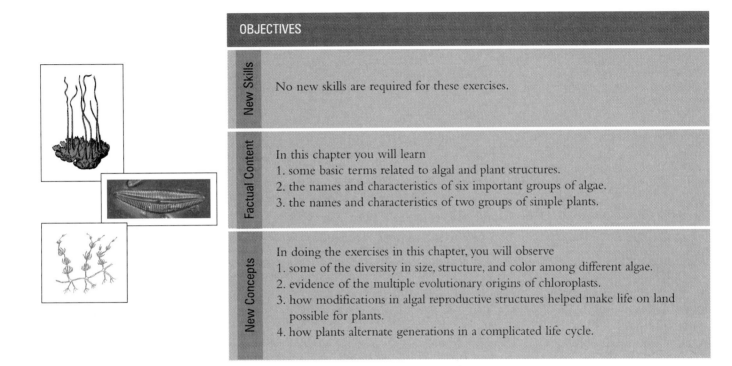

INTRODUCTION

The defining characteristic of eukaryotic cells is the nucleus. Even today there exist several species of unicellular protists that have nuclei but no other organelles. Most eukaryotes, however, have mitochondria. All mitochondria use similar enzymes and have similar nucleic acid sequences, suggesting that they all arose from a single ancestor. That is, shortly after nucleated cells arose, one of them obtained mitochondrial symbionts, and this creature became the ancestor of virtually all of the eukaryotes alive today.

In contrast, many eukaryotes do not have chloroplasts, and those that do have chloroplasts often differ greatly from group to group. Because of this, biologists infer that chloroplasts evolved several times in several different lineages, long after mitochondria were established.

Chloroplasts can differ in many ways. At the molecular level, they can have divergent nucleic acid sequences and enzymes. An important area of variation is in photosynthetic pigments and accessory pigments. In chapter 9 we saw that plants contain two distinct types of chlorophyll, designated *a* and *b*. In nature there are actually three varieties of chlorophyll: *a*, *b*, and *c*. **Chlorophyll *c*** is present in some types of algae. You also saw that plants contain carotenoid pigments. These include orange carotene and yellow xanthophyll. Again, these are only two of many carotenoids, compounds that can occur in a range of colors: red, orange, yellow, and brown. A third class of pigments, the phycobilins, are not so widely distributed but are still important. They can be red or blue.

Chloroplasts also have different types of membrane organizations. Mitochondria and many chloroplasts are bounded by a double membrane. This is interpreted as a vestige of endosymbiosis: the inner membrane is derived from the cell membrane of the original symbiont, while the outer membrane would have been contributed by the host when the symbiont was first engulfed. Some chloroplasts are much more complex, however. They may have three or four membranes. A few have internal structures between the second and third membranes that look like tiny nuclei. Because of this, it is thought that the more complicated chloroplasts are derived from endosymbiotic eukaryotic cells, not bacteria. That is, they are really endosymbionts within endosymbionts.

The notion that chloroplasts were among the last organelles to evolve is interesting. It means that the the first eukaryotes were heterotrophs that lived off prokaryotic producers. Presumably, the first chloroplasts originated when eukaryotes ingested photosynthetic cyanobacteria and, instead of digesting them, maintained them as internal symbionts. In fact, there are many species of fungi and animals that do this today. Lichens are fungi that contain photosynthetic cyanobacteria or algae, and animals such as sponges, corals, and sea squirts can also harbor photosynthetic cells. These are not true chloroplasts, as the photosynthetic symbionts can be removed from their hosts and cultured as free-living bacteria or algae. If, over time, the algae in these animals became totally dependent on their hosts, we would be witnessing the origin of a new type of chloroplast.

KINGDOM PROTISTA

In chapter 17 we said that eukaryotes can be divided into four kingdoms: Protista, Plantae, Fungi, and Animalia. The protist kingdom is at the root of eukaryotic evolution, containing many different lines of descent from various endosymbiotic events. Because of this diversity of origins, many biologists have suggested breaking Protista up into several smaller, more closely related kingdoms, but so far none of these proposals has gained wide acceptance.

A kingdom is a broad category of classification. Biologists organize the members of a kingdom into smaller units called **divisions.** Divisions are further subdivided, as we shall see in chapter 22. Zoologists use the term **phylum** instead of division, and since zoologists were the first people to study protozoa, the term is also used for them. As a result, Kingdom Protista is divided into a mixture of phyla and divisions. We have already encountered several representative protozoa in previous chapters: ciliates (Phylum Ciliophora), amoebas (Phylum Rhizopoda), and flagellates (Phylum Zoomastigophora). In this chapter we will consider several divisions of algae.

ALGAE

The first eukaryotes with chloroplasts were **algae**. Algae are aquatic organisms, living in either fresh water or seawater; a few can live in moist terrestrial environments. The modification of some algae to live on dry land led to the origin of plants.

Algae are a diverse group of organisms. They differ in many ways, not just in their chloroplasts, and can be distinguished by their cell wall materials, internal cell structures, and details in their methods of mitosis. Again, this evidence suggests that these groups had diverged before they acquired chloroplasts. It is impossible to look at all of the various kinds of algae in an introductory course, so we will focus on a few illustrative types.

UNICELLULAR ALGAE

Two widespread groups of unicellular algae are the **diatoms** (Division Bacillariophyta) and **dinoflagellates** (Division Dinoflagellata). In many aquatic ecosystems these creatures are important producers at the base of the food chain. Diatoms are best known for their unique cell walls made of **silica**, the same material found in glass (see figure 18.1). The cell walls of dead diatoms form a gritty material called **diatomaceous earth**, which has many economic uses: as a fine abrasive material, as a reflective glitter for paints, or as a filtration medium. Diatom chloroplasts contain chlorophylls *a* and *c*; carotenoid accessory pigments give them a golden-brown color.

Figure 18.1 Cymbella, *a common diatom*

Dinoflagellates also contain chlorophylls *a* and *c* and carotenoids. As a result they range in color from red to brown. These are the organisms responsible for "red tides," huge algae blooms that

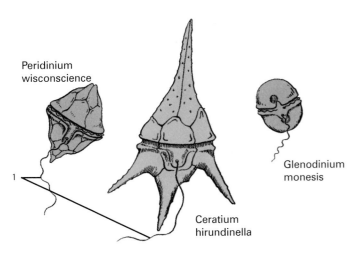

Peridinium
wisconscience

Glenodinium
monesis

Ceratium
hirundinella

Figure 18.2 *Representative dinoflagellates*
1. flagella

Figure 18.3
Rhodomenia, *a*
red alga

Figure 18.4
Sargassum, *a brown alga*
1. air bladder
2. blade
3. stipe

often occur in oceans. Many species of dinoflagellates produce toxic compounds, and red tides generally poison fish and mollusks in great numbers. They can also be a significant health hazard to humans. Dinoflagellates possess a pair of flagella, and some species are surrounded by unusual cell walls made of cellulose plates (figure 18.2).

MACROSCOPIC ALGAE

Some algae are multicellular and can grow to enormous sizes. **Red algae** (Division Rhodophyta) and **brown algae** (Division Phaeophyta) are both multicellular types (some species of red algae are unicellular, but they are not typical). These groups are essentially marine; only a few species of red algae (and no known species of brown algae) live in fresh water. People in coastal areas use both types for food.

Red algae are unusual in many ways. They produce several peculiar carbohydrates, including **agar**, which we use in the laboratory to solidify bacterial media, and **carrageenan**, which is often used to thicken foods like ice cream. Their chloroplasts contain chlorophyll *a* but not chlorophyll *b* or *c*. They are unique among eukaryotes in using phycobilins as accessory pigments. Thus they are usually red in color (figure 18.3) but can also be blue or even black. The pigments found in red algae are apparently very good at scavenging energy from dim light, as these organisms are frequently found deeper in coastal waters than other algae.

Ranging in length from 3 inches to 300 feet, the brown algae include the largest species of algae. The most complex are commonly called "kelps" or "seaweed," and they are often differentiated into specialized organs (figure 18.4). Their chloroplasts contain chlorophylls *a* and *c* and various carotenoids.

THE ANCESTORS OF PLANTS

Common plants are not like any of the algae considered so far. As you know, their chloroplasts contain different pigments; there are also many structural, chemical, and genetic differences among the organisms as a whole.

There are two groups of algae that are very similar to plants: the **green algae** (Division Chlorophyta) and the **charophytes** (Division Charophyta). Both use chlorophylls *a* and *b*, store carbohydrates as starch, and make their cell walls out of cellulose. Green algae are incredibly diverse. Freshwater or marine, they can range from unicellular to multicellular and come in all shapes and sizes. This diversity suggests that green algae are an ancient group that has had a great deal of time to adapt to many environments and ways of life.

Like most protists, the greens are capable of sexual reproduction. In the simpler types this is often by a form of conjugation. Unlike ciliate conjugation, however, conjugation in green algae is not limited to the exchange of micronuclei, but it also involves the fusion of two haploid cells. This is accomplished by fusing the cell walls of male and female cells so that the male cell can crawl inside the cell wall of the female and join with it (figure 18.5). The **zygote**, or fertilized egg, develops in the old cell wall of the egg cell.

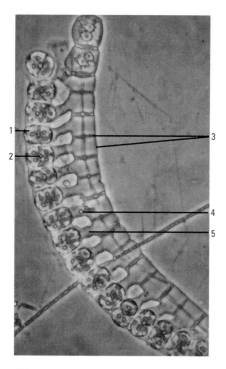

Figure 18.5 Spirogyra *undergoing conjugation*
1. zygote
2. chloroplast
3. cell wall
4. male cell
5. conjugation tube

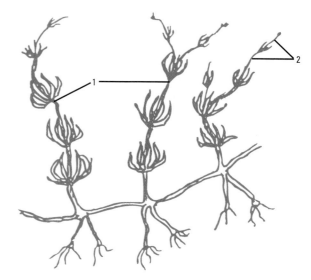

Figure 18.6 Chara *morphology*
1. nodes
2. internodes

The charophytes are a younger group than the green algae. They differ from the greens in a variety of cellular and molecular characteristics. There are several traits found in plants that seem to have their origin in the charophytes. One charophyte, whose scientific name is *Chara* (the whole group is named after this alga), is especially interesting in relation to plant evolution.

The body of *Chara* is quite different from any green alga in that it is organized into **nodes** and **internodes**. Nodes are regions of the organism where branching occurs, and internodes are the long straight sections between the nodes (figure 18.6). Plants are arranged in the same way: their stems and branches are internodes and the points where branches or leaves occur are nodes. This structure results from a special pattern of growth called **apical growth**, a pattern *Chara* also shares with plants. "Apical growth" describes the fact that new growth in plants occurs only at the very tips of the stems, branches, and roots. The part of the plant nearest the ground is the oldest and never grows longer once it is mature. New tissues are formed by cell division, and this only takes place at the apexes of the shoots. These areas of mitosis are called **apical meristems** (from the Greek *merizo*, to divide). Nodes, internodes, and apical growth explain why plants look the way they do – thin and branching rather than having a concise and symmetrical form like animals.

Chara shares another trait with true plants: the presence of sterile, protective cells around the sperm and eggs. In algae a chamber that holds an egg is called an **oogonium** (from the Greek *oon*, "egg"), and a chamber that holds sperm is called an **antheridium** (from the Greek *anthos*, "flower," the reproductive organ of many plants). As shown in figure 18.5, the oogonia and antheridia of green algae (and some charophytes) are simply the cell walls of the mother cells that produced the sperm and eggs. However, in *Chara* the oogonia are surrounded by spiral filaments of nonreproductive cells, cells that cannot by themselves produce eggs (figure 18.7). What this extra layer of protection does for the charophytes is not clear – most algae are able to survive quite well without it. On the other hand, it turns out to be crucial for land plants. Even in the simplest land plants this extra layer of cells has developed into a structure that not only protects the egg but also holds the zygote after fertilization and keeps the young embryo from drying out.

This is an interesting illustration of an important concept in evolution. It is apparent that this protective layer around the oogonium arose before it was most needed (that is, in aquatic charophytes). Traits arise out of the natural processes of variation we studied in chapters 13 and 14, and then these new and different traits become the basis for new adaptations. There is no way an organism in a new environment can suddenly create the adaptations it needs. It will only survive if it already has them in at least a rudimentary form. The presence of these variations is what allows the organism to successfully exploit the new environment. Structures that arise in one environment but become the basis for adaptations to another environment are often called **preadaptations.** Despite the name, preadaptations do not imply any kind of "planning" by an organism. Rather, they are simply existing structures that are co-opted for some new use. (Some

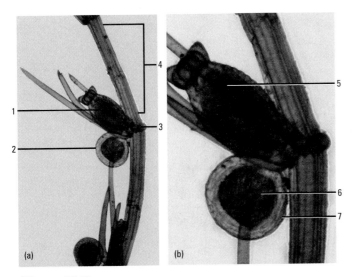

Figure 18.7

(a) Chara, *sexual structures. (b) magnified view*
1. *oogonium*
2. *antheridium*
3. *node*
4. *internode*
5. *oogonium*
6. *sperm*
7. *antheridium*

biologists have proposed changing the name of this term to *exaptation* or *coaptation*, but so far these terms are not widely used.)

It must be stressed that *Chara* is a modern organism and not the ancestor of the plant kingdom. In fact, it does not have all of the traits that the ancestral plant must have had. Another key trait, seen in a different charophyte called *Coleochaete*, is the presence of a waxy coating called a **cuticle**. A plant requires a cuticle to retain moisture; if it is washed off with nonpolar solvents, the plant will quickly wilt and die. Since *Coleochaete* is an aquatic organism, its cuticle performs some other function (it may prevent other algae from growing on top of it and blocking its light). Thus the cuticle is another example of a preadaptation.

THE SIMPLEST PLANTS

Ecologically, the major difference between plants and algae is that plants live on land. As we have seen, the move onto land was made possible by characteristics found in the aquatic charophytes. Presumably, the ancestor of plants had a body with nodes and internodes and complex oogonia like *Chara* and a waxy cuticle like *Coleochaete*. It seems likely that this creature lived in some kind of shoreline habitat that alternated between wet and dry, and some of its offspring were able to further exploit the dry environments further from shore.

Water plays a key role in sexual reproduction, as sperm cells must be able to swim to eggs to create zygotes. Terrestrial animals deal with this problem through mechanisms by which males deliver sperm to the eggs during copulation, but this is not an option for plants rooted in place. Instead, plants use a complex life cycle of alternating generations to make sure that sexual reproduction always takes place in a moist environment.

Alternation of generations is another trait that is found in various types of algae (it occurs in certain species of red, brown, and green algae and seems to have arisen in different ways in different lineages at different times). When generations alternate, a diploid individual gives rise to a haploid individual and a haploid individual gives rise to a diploid individual. This is quite different from the way animals reproduce. Virtually all animals are diploid, as are their offspring; the only haploid cells in the average animal life cycle are sperm and eggs.

If you were the product of alternation of generations, your family would be very different. Your mother might still be a normal diploid person, producing normal, haploid reproductive cells. However, instead of being fertilized, one of those cells would have grown directly into you! All of the cells in your body would therefore be haploid. You would have no father. Genetically, you would only be half of what your mother was. You would not require meiosis in order to make sperm or eggs, as all of your cells would already be haploid. You would mate with another haploid individual but then have diploid children. Your mother would not see her diploid chromosome number restored until her grandchildren were born, and you would not see another haploid generation until your own grandchildren were born. The two chromosome numbers alternate. Strange as this sounds when applied to humans, it is normal in plants and many algae.

In some algae the haploid and diploid organisms look very much alike. This is not so in plants, Each is specialized for a particular role. Haploid plants are generally small and limited to moist environments — that way there is water for the sperm to swim in and the distance from male plant to female plant is small. Diploid plants make asexual spores instead of sperm and eggs. These spores have thick walls to protect them from dehydration and are dispersed by wind. Spores are haploid (they are produced by meiosis), and those that land in a suitable environment germinate into haploid plants. Thus the diploid plants can live in relatively dry environments as long as they are able to disperse their spores to a nearby moist habitat so the haploid generation can grow. Diploid plants are called **sporophytes** because they produce spores. Similarly, haploid plants are called **gametophytes**. Sporophytes may be either small or large (grass and trees are both sporophytes) but gametophytes are always small.

The plant kingdom is partitioned into several divisions. The simplest plants, hornworts and mosses, belong to Divisions

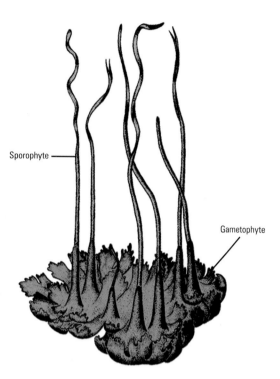

Figure 18.8 *A hornwort*

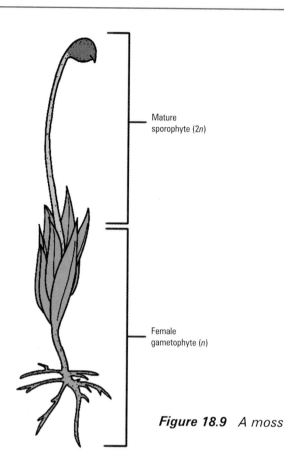

Figure 18.9 *A moss*

Anthoceratophyta and Bryophyta, respectively. The sporophytes of hornworts and mosses are small. In fact, these are the only plants whose sporophytes are smaller than their gametophytes. The gametophytes of hornworts are small green lumps. They develop by apical growth like other plants, but are so compact and undifferentiated that they do not exhibit clear nodes and internodes. The females produce eggs inside structures analogous to the oogonia of *Chara*, except that they are more elaborate and are therefore given another name: **archegonia**. Since an archegonium is not under water, it cannot simply release a zygote to float away, but must instead support the zygote after fertilization and actually ends up holding the new sporophyte plant like a vase. Male plants produce sperm inside antheridia just like *Chara*.

Young plant sporophytes are always rooted in their gametophyte mothers. If the sporophyte grows much larger than the gametophyte, it eventually destroys the gametophyte. On the other hand, if the gametophyte is larger than the sporophyte, as it is in hornworts and mosses, the two live together harmoniously. The archegonia of hornworts are embedded in the upper surface of the gametophyte bodies. When a hornwort sporophyte grows out of an archegonium it looks like it is growing on top of the gametophyte. The sporophyte is tall and slender and looks like a horn (figure 18.8), hence their common name (*wort* is an old Anglo-Saxon word for herb). Cells inside the sporophyte undergo meiosis and fill the plant with spores. Eventually it bursts and

releases the spores to begin a new generation of gametophytes.

Mosses are more complex than hornworts, and some species can grow to be over a foot tall. The moss gametophyte consists of a slender shoot with "leafy" outgrowths that are too simple to be considered true leaves. At the tops of mature plants, archegonia or antheridia form. The sperm are released amidst rain or heavy dew to swim toward an archegonium; occasionally raindrops will splash the sperm directly onto the archegonium and shorten the journey. The fertilized zygote develops in a manner similar to that of a hornwort: a long stalk growing on top of the established gametophyte (figure 18.9). When the sporophyte reaches its mature length, it develops an enlarged head full of haploid spores, which are released when the head breaks open.

EXERCISES WITH ALGAE AND SIMPLE PLANTS

Most of the following exercises are relatively descriptive – instead of performing tests to see what happens in a particular situation, you will make observations with the goal of learning how different organisms are put together and how they resemble or differ from each other. Sketch what you see and answer the associated questions based on your observations.

A. Algal diversity

1. **Dinoflagellates**: Examine a prepared slide of *Ceratium*. The cell wall of this organism has several long, thin arms that give it a shape reminiscent of the Eiffel Tower in Paris. See if you can locate the groove around the middle of the organism. What color are the chloroplasts?

2. **Diatoms**: Examine a prepared slide of diatoms. All you are actually seeing are the leftover cell walls of dead diatoms—note how they refract light like cut glass. Why is that?

3. **Red algae**: Observe the range of body types and sizes displayed on the herbarium sheets of dried red algae. How do the colors compare to those of land plants?

4. **Brown algae**: Examine the morphology of preserved specimens on herbarium sheets. Note the range of colors. Can you find specialized body regions as depicted in figure 18.4?

B. Chloroplast diversity and origins

As you saw in chapter 9, the pigments in chloroplasts can be extracted, separated by chromatography, and analyzed by spectrophotometry. Scientists have done this for many types of algae and bacteria, and the results for a few are summarized in the following table. The table also indicates how many membranes bound the chloroplasts, as indicated by electron microscopy.

Organism	Chlorophylls	Membranes	Accessory pigments
cyanobacteria	*a*	1	carotenoids & phycobilins
brown algae	*a & c*	4	carotenoids
charophytes	*a & b*	2	carotenoids
diatoms	*a & c*	4	carotenoids
dinoflagellates	*a & c*	3 or 4	carotenoids
green algae	*a & b*	2	carotenoids
red algae	*a*	2	carotenoids & phycobilins

As you know, carotenoids are actually a whole class of pigments that can be separated into individual compounds (such as carotene and xanthophyll) by chromatography. The same is true

of phycobilins. We will not concern ourselves with these details here – instead we will treat each class as a single pigment.

1. Divide the organisms in the table into groups having similar pigments. Where does the prokaryote fit best? What modifications does it need to fall in with the other types of algae?

2. Construct a simple tree showing how an ancestral eukaryotic line could have branched into the groups identified in step 1. Does each branch suggests a separate endosymbiotic origin for the chloroplasts in that group? Do any of the branches then sub-divide after the chloroplasts are acquired? What is the minimum number of endosymbiotic events necessary to account for these data? Can you tell if history might have entailed more than this minimum number? What type of alga appears to have had two separate origins?

C. Important algal structures

1. **Morphology**: Examine a preserved specimen of *Chara*. Identify the nodes and internodes and compare the algal body to that of more familiar plants.

2. **Oogonia**: Examine a prepared slide of *Spirogyra* conjugation and look for oogonia containing zygotes. Then look at a slide of *Chara* oogonia and study the spiral filaments surrounding the egg. How does this compare to *Spirogyra*?

D. Land plant structures

1. **Hornworts**: Study the preserved specimens of the hornwort *Anthoceros*. Identify the sporophyte and the gametophyte. Then look at a prepared slide of a sporophyte and identify the spores inside.

2. **Mosses**: Observe the preserved specimens of moss and identify the sporophytes attached to the gametophytes. Have you ever observed either of these growing outdoors? Which one — sporophytes or gametophytes? How do the moss gametophytes compare to the hornwort gametophytes? How do the sporophytes compare?

IMPORTANT TERMS

agar
algae
alternation of generations
antheridium
apical growth
apical meristem
archegonium

brown algae
carrageenan
charophyte
chlorophyll *c*
cuticle
diatom
diatomaceous earth

dinoflagellate
division
gametophyte
green algae
internode
node
oogonium

phylum
preadaptation
red algae
silica
sporophyte
zygote

Laboratory Report Algae and the Origin of Plants

FACTUAL CONTENT

Define the following terms. Brief definitions are given in the chapter, but you may also wish to consult your lecture notes and text for a better idea of how the terms are used. Pay close attention to those words that look similar but have different meanings — you will want to be able to tell them apart if they appear next to each other on a multiple-choice exam.

agar

algae

alternation of generations

antheridium

apical growth

apical meristem

archegonium

brown algae

carrageenan

charophyte

chlorophyll *c*

cuticle

diatom

diatomaceous earth

dinoflagellate

division

gametophyte

green algae

internode

node

oogonium

phylum

preadaptation

red algae

silica

sporophyte

zygote

After the common name of each group of organism, write the formal scientific name of the corresponding division:

brown algae

charophyte

diatom

dinoflagellate

green algae

red algae

hornworts

mosses

EXERCISES - Record your data and answer the questions below.

A. Algal diversity

1. **Dinoflagellates**: Sketch *Ceratium*:

Label the groove around the middle of the organism.

What color are the chloroplasts?

2.   **Diatoms**: Sketch the diatom cell walls:

Why do they appear glassy?

3.   **Red algae**: Sketch or describe the red algae on display:

How do the colors compare to land plants?

4. **Brown algae**: Sketch or describe the brown algae on display:

Can you find specialized body regions as depicted in figure 18.4?

B. Chloroplast diversity and origins

Divide the organisms in the table into groups having similar pigments:

Organism	chlorophylls	accessory pigments	membranes

Where does the prokaryote fit best? What modifications does it need to fall in with the other types of algae?

Construct a simple tree showing how an ancestral eukaryotic line could have branched into the groups identified in step one:

Does each branch suggests a separate endosymbiotic origin for the chloroplasts in that group?

Do any of the branches then sub-divide after the chloroplasts are acquired?

What is the minimum number of endosymbiotic events necessary to account for these data?

Can you tell if history might have entailed more than this minimum number?

What type of alga appears to have had two separate origins?

C. Important algal structures

1. **Morphology**: Sketch a preserved specimen of *Chara*:

Label the nodes and internodes.

2. **Oogonia**: Sketch *Spirogyra* conjugation:

Label oogonia and zygotes.

Sketch *Chara* oogonia:

Label the spiral filaments surrounding the egg.

How does this compare to *Spirogyra*?

D. Land plant structures

1. **Hornworts**: Sketch a preserved specimen of the hornwort *Anthoceros*:

Label the sporophyte and the gametophyte.

Sketch a prepared slide of a sporophyte:

Label the spores inside.

2.   **Mosses**:   Sketch the preserved specimens of moss:

Label the sporophytes and the gametophytes.

Have you ever observed either of these growing outdoors? Which one – sporophytes or gametophytes?

How do the moss gametophytes compare to the hornwort gametophytes?

How do the sporophytes compare?

The Rise of the Animal Kingdom

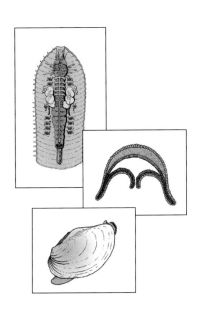

INTRODUCTION

Once eukaryotes first appeared on earth, algae and animals followed. The oldest animal fossils known are roughly 800 million years old; they show impressions of strange creatures that are very different from the animals familiar to us today. No one knows how they are related to modern forms, if, indeed, they are related at all. Later fossils, 500 to 600 million years old, show an assemblage of different creatures that rapidly diversified. In this second group of animals we can recognize all of the basic body plans in the modern animal kingdom, as well as numerous species built on body plans that are now completely extinct. All animals alive today are merely variations of those body types that have survived from 500 million years ago. Each body plan defines a different phylum of the animal kingdom.

It is a little misleading to refer to those "animals familiar to us today." Most modern animals are not very familiar to many people. That is because humans are terrestrial creatures, whereas the vast majority of animals are, like algae, aquatic and generally marine. In this chapter we will survey several common phyla of animals, but only the last two contain species that you are likely to have seen regularly. The animals we find most familiar belong to groups that we will not consider until chapter 21.

WHAT IS AN ANIMAL?

Like protozoa, animals are heterotrophic. They are usually holotrophs, ingesting their food rather than absorbing it, although many animals live parasitically from the body fluids of other animals.

Unlike protozoa, animals are multicellular, and this means that they can bring many new adaptations to a holotrophic existence. For one thing, they can grow much larger than a single-celled organism and therefore can potentially exploit a wider range of possible foods. An amoeba mainly eats bacteria, but huge whales have more choices. The presence of many cells also allows for tissue specialization. Animals (except a few highly specialized parasites) have mouths and digestive systems. Most have muscle cells that permit movement and nerve cells to carry messages from one part of the body to another quickly. Coordinated movement is a great advantage in finding and capturing food.

An animal begins life as a single cell – a zygote. The zygote divides by mitosis to produce a multicellular body, and the various cells specialize as they appear. They do not follow a pattern of apical growth as we saw in *Chara* and land plants, and so they do not develop a body of nodes and internodes. Instead, animals exhibit very intricate patterns of development as they grow from zygotes into adults. The study of an individual organism's development is called **embryology**. Embryology provides many clues about the evolutionary relationships between animals. In many cases, the difference between one type of animal and another is due to a small shift in developmental patterns early in life. As we shall see later in this chapter, creatures that are virtually identical at one stage of development may end up looking completely different as adults.

As before, we must limit the scope of our survey of diversity to a few representative types. The groups of animals described, while representing a wide array of body plans, exhibit structural and embryological features that suggest evolutionary relationships. While future discoveries may require changes in our ideas of how animals evolved, present evidence indicates that this line of descent includes the bulk of the animal kingdom as we know it.

CNIDARIANS
Kingdom Animalia, Phylum Cnidaria

Animals achieve complexity by differentiating their cells; the less cell differentiation they undergo, the simpler they are. Thus, the simplest shape an animal can have is roundness: being the same in all directions. Since an animal must have a mouth, the perfect symmetry of a sphere is impossible, as the mouth would be on one side only. Therefore the simplest shape available to animals is **radial symmetry**, that is, round with a definite top and bottom (like a cylindrical vase with an opening in the top). And, in fact, this is exactly the shape we see in a large group of animals whose bodies are made of very few kinds of cells. These animals are called **cnidarians** (from the Greek *knide*, nettle, a prickly plant, which refers to the stingers they have on their tentacles).

Cnidarians are widespread in oceans and to a much lesser extent in freshwater. One group, the **corals**, is very important ecologically because they build **coral reefs**, which serve as habitats for many kinds of animals and algae. Cnidarians frequently become infested with photosynthetic algae. While this can be an important factor in the food chain of a coral reef, it also makes the reef very sensitive to light and other factors that affect algal growth.

The body of a cnidarian is essentially a sack with a single opening – the **mouth**. Surrounding the mouth is a ring of **tentacles** equipped with stinging cells called **cnidocytes**. Cnidarians use the tentacles to grab prey and force it into their mouths. The body wall of a cnidarian is only two cells thick, although the two layers are separated by a gelatinous matrix that can contribute additional bulk. The outer layer is the **epidermis** and the inner lines the **gastrovascular cavity**, or digestive chamber. Because of the gastrovascular cavity, cnidarians can ingest food particles larger than single cells. Anything that is not digested is regurgitated out the mouth, as the animals have no anus. Some of the cells of the body wall are specialized as nerve or muscle cells, but these are scattered throughout the body in a loose network rather than being organized into distinct organs. Therefore cnidarians can move about but exhibit little coordination.

This basic body plan can take two forms: **polyps**, characteristic of anemones and corals, and floating **medusae**, exemplified by **jellyfish** (figure 19.1). Most cnidarians have a life cycle that alternates between the two shapes. It is important to remember that animal life cycles are not the same as plant life cycles – animals do not alternate between haploid and diploid individuals. The only haploid cells in a typical animal life cycle are sperm and eggs. An animal life cycle shows how a zygote develops into a larva and then into successive adult shapes, all of which are diploid.

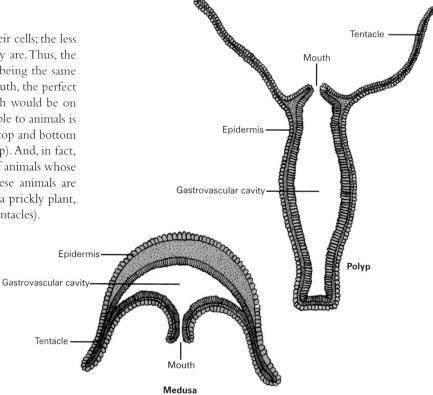

Figure 19.1 Cnidarian body types

Figure 19.2 illustrates the life cycle of the cnidarian *Obelia*. The zygote grows into a hollow ball called a **blastula**, a developmental stage found in all animals (except sponges). In cnidarians the blastula fills with cells and becomes a solid larva called a **planula**. Planulae are covered with microscopic hairs called **cilia**. Cilia are similar to flagella and can be used to propel cells through water; cilia make the planula free-swimming. The planula eventually attaches itself to a suitable substrate and grows into a polyp, forming a mouth, a gastrovascular cavity, and tentacles. *Obelia* produces long branching colonies of polyps. Some of the polyps become specialized to generate buds of tissue that break off from the colony and develop into independent medusae. The medusae are the sexually mature adults, producing sperm and eggs that can fuse into new zygotes and complete the life cycle.

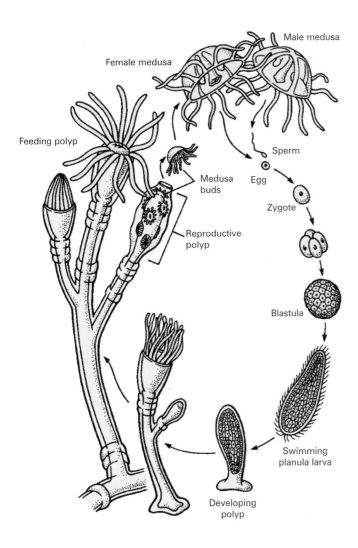

Figure 19.2 *Life cycle of* Obelia, *a colonial cnidarian*

Labels in figure:
Male medusa
Female medusa
Feeding polyp
Sperm
Medusa buds
Egg
Zygote
Reproductive polyp
Blastula
Swimming planula larva
Developing polyp

FLATWORMS
Kingdom Animalia, Phylum Platyhelminthes

Cnidarians are effective predators and, as a group, have survived for half a billion years. Nevertheless they are limited by their structural simplicity.

Animals searching for food are helped by a sense of direction. This requires stretching the animal into a **bilaterally symmetrical** shape, one that not only has a top and bottom but also a front and back and left and right. This kind of shape is found in virtually all animals, from various kinds of worms to large vertebrates. In most cases the front end of the animal exhibits a concentration of sense organs and nerve tissue to interpret the input from these organs – that is, it has a head and a **brain**. Although brains and sense organs can be modified to perform other functions in addition to finding food, the ability to seek nutrients is built into their basic structure. Even in a human brain the chemicals that transmit signals from one nerve to another are mostly made out of amino acids, the very nutrients that are most critical to an animal's diet. (Experiments have shown that insects and other creatures instinctively choose diets that maximize these amino acids.)

Zoologists use specific terms to describe the orientation of a bilateral body: **Anterior** refers to the head end of the animal, while **posterior** refers to the tail. An animal's belly is on the **ventral** side, and the back is **dorsal**. These terms are unambiguous because they are independent of the way an animal happens to be positioned – for example, the spines of both dogs and humans are dorsal, even though the dog spine is horizontal and the human spine vertical.

The simplest bilaterally symmetrical animals are **flatworms**. Flatworms are creatures that look something like elongated planulae with gastrovascular cavities. In fact, it is possible that they evolved from planulae that, due to some shift in their embryological patterns, developed gastrovascular cavities as larvae instead of first transforming into polyps. Most flatworms have heads, although this feature can be reduced or absent in those species that are adapted for a parasitic lifestyle. Free-living flatworms are found in marine and freshwater habitats and occasionally in the moist soil of tropical environments. Parasitic flatworms include **flukes**, **schistosomes**, and **tapeworms**.

Like cnidarians, flatworms have a single opening into the gastrovascular cavity. The gastrovascular cavity tends to be highly branched so that it reaches into all portions of the animal; flatforms have no circulatory system to distribute materials throughout their bodies. Other organs may be present, depending on what type of environment the species is adapted for.

Figure 19.3 is an illustration of the anatomy of the common free-living flatworm, **planaria**. Planarians are active predators and

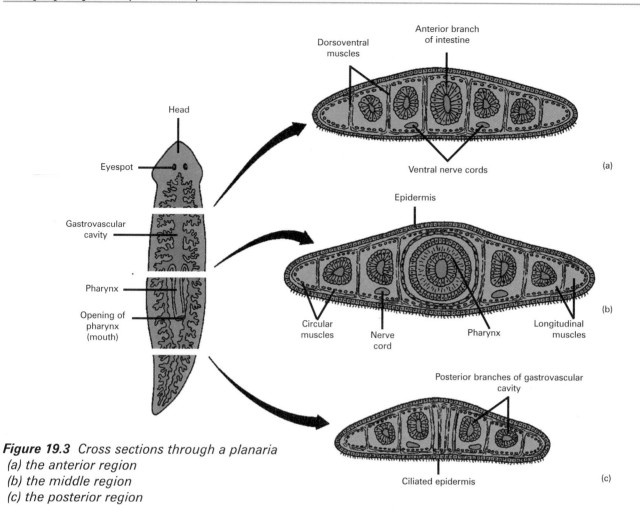

Figure 19.3 *Cross sections through a planaria*
(a) the anterior region
(b) the middle region
(c) the posterior region

scavengers. Their heads have light-sensitive **eyespots** and other sense organs. The planarian nervous system is well developed; planarians can even be trained to follow simple paths and respond to particular stimuli. A planarian's mouth is not in its head; it is at the end of a muscular tube (the **pharynx**) that extends from the gastrovascular cavity out the ventral side of the worm. As the worm crawls about its environment (like a planula, it is covered with cilia), the pharynx sucks up food like a vacuum hose.

ROUNDWORMS
Kingdom Animalia, Phylum Nematoda

The **roundworm** body plan has two features not found in flatworms. First of all, the tissue surrounding their digestive tracts is not solid and fleshy – the internal organs are suspended in a fluid-filled space called a **pseudocoelom**. Secondly, roundworms have a complete digestive system with two openings, a mouth and an **anus**. Because of these two features, their body plan can be thought of as a tube within a tube. From now on, all the animals we will study will have tube-within-a-tube body plans. The presence of an internal body cavity allows animals to develop many specialized organs without sacrificing physical flexibility and movement, while a complete digestive tract means

that digestion occurs in an efficient, one-way "assembly line." It is a pattern that can be modified into a wide variety of forms.

Figure 19.4 illustrates the anatomy of *Ascaris*, a kind of roundworm that lives parasitically inside the intestines of other animals. The tube-within-a-tube body plan is evident. Much of the pseudocoelom space is taken up by reproductive organs.

There are nearly 100,000 known species of roundworms, but zoologists believe that number is only a fraction of the species that probably exist. Roundworms are plentiful in damp soil and aquatic habitats. They also parasitize numerous plants and animals. Humans are affected by many types of roundworms, including hookworms, ascarids, and *Trichinella*, a parasite carried in pork. Roundworms can range in size from a few millimeters to several centimeters.

SEGMENTED WORMS
Kingdom Animalia, Phylum Annelida

At first glance, **segmented worms** (or **annelids**) look very much like roundworms and appear in the same general size range. Their body plan has a pair of subtle differences, however.

The tissues lining the body cavity of a segmented worm are arranged differently than those of a roundworm, and so the cavity is considered a true **coelom** instead of a pseudocoelom. And, as the name of the group implies, their body walls are arranged in repetitive **segments**. Segmentation is basically limited to the body wall and its musculature – the digestive tract and many other organs extend through multiple segments and are not repeated. Segmentation makes a big difference in how the worms move and also make possible, in some marine worms, a system of repetitive appendages that can be adapted to many uses.

Like the other types of worms, segmented worms are generally aquatic, with a few species living in moist terrestrial habitats. The best known segmented worms are leeches and common earthworms.

The internal anatomy of an earthworm is illustrated in figure 19.5. It is roughly similar to that of a roundworm, a long tube within a tube. In the anterior segments the inner tube, the digestive tract, is differentiated into several organs. The muscular **pharynx** helps to draw food into the mouth and push it through the **esophagus** to an enlarged storage organ, the **crop**. The crop slowly releases the food to the **gizzard**, which is heavily muscled and grinds the food by mixing it with grit and sand ingested as the worm eats. The resulting mash passes into the long **intestine** that extends through most of the worm. Here enzymes digest the food into small molecules that can be absorbed into the blood. A large dorsal blood vessel runs down the back of the digestive tract and moves blood past the intestine continuously. A second blood vessel runs ventral to the digestive tract and is connected to the dorsal vessel by five simple hearts in the esophageal region. The **nerve cord** is ventral to the digestive system and swells to a small brain just inside the mouth. The **gonads**, which produce sperm and eggs, are also located in the anterior end of the worm.

Marine annelids are generally much more complex than earthworms. They have well-developed heads with sense organs and often tentacles. The "tube worms" sold in pet stores for salt water aquaria are segmented worms that hide most of their bodies inside a buried tube, extending only their tentacles and **gills** in a colorful, feathery display.

Marine annelids have pairs of fleshy appendages known as **parapodia** on their segments (figure 19.6). Parapodia can vary greatly in form from species to species, often being modified for tasks such as burrowing and swimming. Parapodia have rich blood supplies, and in some species the fleshier portions of the parapodia act as gills. Anchored in the parapodia are bundles of chitinous bristles, which stiffen the appendages and facilitate movement (see figure 19.7).

MOLLUSKS
Kingdom Animalia, Phylum Mollusca

The segmented worms are part of an important evolutionary line that we will return to in chapter 21. An offshoot of this line leads to a very large and important group of animals, the **mollusks**. Mollusks are unusual in many ways, so we will not dwell too long on their structure, but it is worthwhile to note what they are and how they relate to other animals.

Mollusks are fleshy animals that usually have shells. The group contains thousands of species of clams, snails, slugs, squids, and octopuses (figure 19.8). As with other animals, most are aquatic, with some

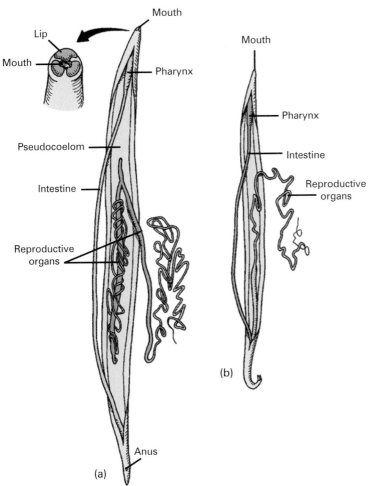

Figure 19.4 *Anatomy of (a) a female and (b) a male* Ascaris

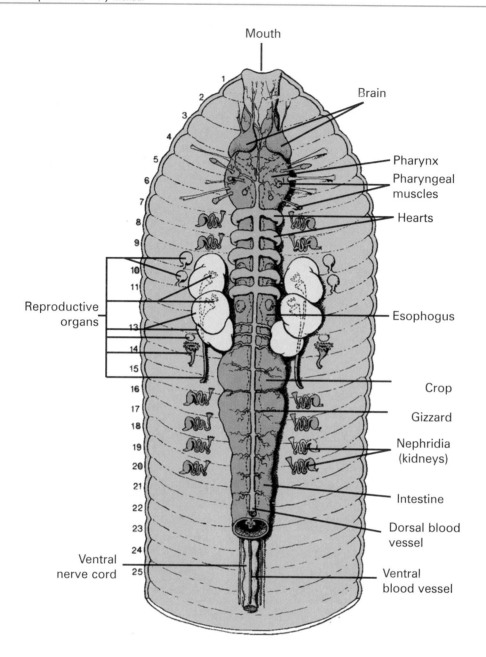

Figure 19.5 *Anatomy of the anterior end of an earthworm*

snails and slugs living in damp habitats on land. The basic mollusk body plan consists of a fleshy body (containing the internal organs) from which extends a muscular **foot**. The body is often shrouded in a hard shell with only the foot exposed. An odd feature of mollusk anatomy is that those species that have heads (clams, for example, do not) have them as regions of their feet. This is readily seen in snails, where the head is the anterior end of the foot that reaches out of the shell. Likewise, the eyes and mouths of squids and octopuses occur near their tentacles (which are simply subdivisions of the mollusk foot), while the main body of the animal is dorsal to this "head-foot" region.

Mollusks are not segmented and bear no obvious resemblance to segmented worms. Surprisingly, embryology shows that they are nevertheless related. The zygotes of mollusks and marine annelids follow very similar developmental paths until they reach a tiny larval stage called a **trochophore**. Trochophores have bands of cilia and swim about as tiny, free-living larvae while they feed and grow. If the trochophore expands in all dimensions, the result is a mollusk (figure 19.9). On the other hand, if the trochophore grows by adding segments at one end, the result is a worm. Thus, the difference between a mollusk and a worm is simply which path of development the animal follows after the larval stage.

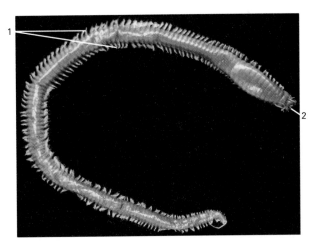

Figure 19.6 *A dorsal view of the sandworm,* Neanthes
1. parapodia   2. mouth

Figure 19.7 *Parapodium of the sandworm,* Neanthes
1. fleshy portion (gill) 2. chitinous bristles

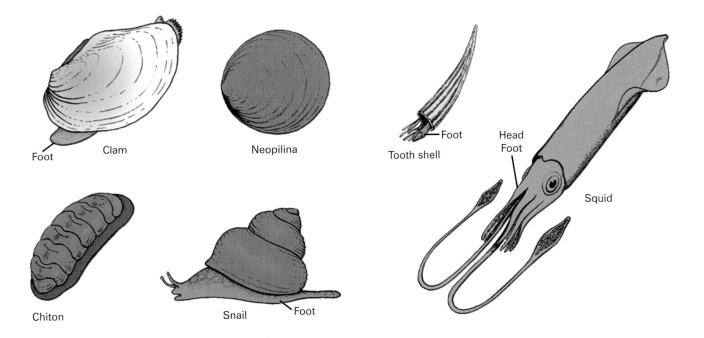

Foot
Clam
Neopilina
Foot
Tooth shell
Head
Foot
Squid

Chiton
Snail   Foot

Figure 19.8 *Some representative mollusks*

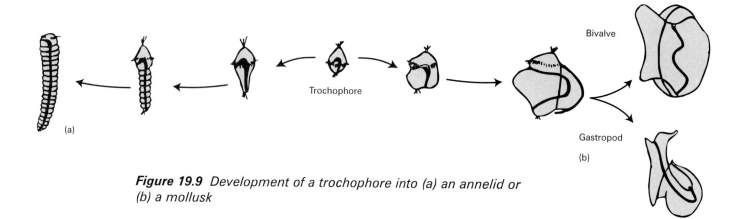

Bivalve

Trochophore

Gastropod

(a)

(b)

Figure 19.9 *Development of a trochophore into (a) an annelid or (b) a mollusk*

EXERCISES WITH SIMPLE ANIMALS

In the following exercises you will work with several living animals. Treat them with care, for they are small and delicate. They require cool, moist surroundings, so do not expose them to any more light than is necessary for your observations, and never let them dry out. You should be able to see them move and explore their environment as they search for food.

A. Cnidarian structure and behavior

1. **Hydra**: Place some living hydra in a deep well slide or shallow dish. Observe their motion and how they can change their body shape. Do they show evidence of very coordinated movement? Depending on availability, some of the hydra may have green algae growing symbiotically in their tissues. Try to find some of these specimens.

2. **Cnidarian colonies**: Examine a prepared slide of an *Obelia* colony. How does the colony compare to a single hydra? Refer to figure 19.2. Can you distinguish the feeding polyps from those specialized for producing medusae?

3. **Cnidarian medusae**: Examine a prepared slide of an *Obelia* medusa. How does it compare to a polyp?

4. **Planula larvae**: Examine a prepared slide of a cnidarian planula. Sketch what you see so you can compare it to a planarian later.

B. Flatworm structure and behavior

1. **Planarian morphology**: Examine a prepared slide of a planaria and locate the gastrovascular cavity and pharynx (see figure 19.3). How does the gastrovascular cavity compare to that of a hydra?

2. **Planarian behavior**: Place a living planaria in a deep well slide or shallow dish. Do you see a head with eyespots? When it moves, what direction does it go? How does it respond to light? With high enough magnification you may be able to see the beating cilia on the surface of its body. If some food (cooked egg yolk, etc.) is available, put some in the well and see if the worm will feed.

C. Roundworm dissection

Ascarids are relatively large, parasitic roundworms that live in vertebrate intestines. Their size makes them reasonably easy to dissect.

1. Obtain a preserved *Ascaris* worm and place it in a dissecting pan. Carefully cut through the body wall with a scalpel and slice it from end to end. Because of the pseudocoelom, it should open up to reveal an intestine and reproductive organs suspended in an internal cavity.

2. Trace the digestive system and observe the tube-within-a-tube body plan. Draw what you see and compare it to figure 19.4.

D. Earthworm dissection

Compared to roundworms, earthworms are fairly complex animals. For this dissection you will need to work slowly and carefully in order to expose all of the listed organs. Once a structure is cut apart, it cannot be put back together again! Use forceps to manipulate the structures and expose deeper tissues, cutting only when necessary. Sometimes it will be necessary to use a scalpel to cut into a surface, other times it will be easier to cut with scissors. When cutting, you will find your movements easier to control if you cut away from you while resting your elbows on the table. Use pins to probe small areas and to pin open incisions. When pinning your specimen to your tray, position it near one edge of the tray so that you can view it under the objective of a dissecting microscope. The microscope will be crucial for seeing details.

1. Obtain a preserved earthworm from your lab instructor. Before pinning it down, look at some of the external features. Can you see the mouth and anus? Rub your finger along the ventral surface and feel the tiny bristles. How might these help the worm move? Put the worm, ventral side down, in your dissecting pan and pin down the head and tail.

2. Make an incision down the dorsal surface, taking care not to cut too deep. Carefully run a pin along the inside of the skin on each side of the incision to destroy the internal segmental membranes – this will make it much easier to open the incision and expose the viscera.

3. Open the incision and pin the flaps of skin to each side. The gonads and associated structures should be visible as small, whitish particles packed along the digestive tract in the anterior segments. Gently push them aside with a pin.

4. Starting at the mouth, locate the major sections of the digestive system illustrated in figure 19.5. Locate the dorsal blood vessel and the hearts. Move the intestine aside and look for the ventral nerve cord. If you have been extremely careful, you may have preserved the tiny brain anterior to the pharynx. Look for it under the dissecting microscope. Draw what you see in your notes.

E. Parapodia

1. Examine the various preserved worms on display and see the different types of parapodia on the segments.

2. Study the parapodia in more detail on a prepared slide. Compare what you see to figure 19.7. Can you find the chitinous bristles?

F. Mollusk diversity

1. **Snail behavior**: Watch some snails in an aquarium tank. Locate the various body regions. Can you see the head portion of the foot? Where is the mouth?

2. **Mollusk diversity**: Several preserved mollusks are on display. Locate the shell (if present) and the foot and body regions of the basic mollusk body plan in each specimen. How are these modified in each animal?

G. Body plans

1. Construct a table to compare body plans. In the first column list each of the five phyla you have studied.

2. Make columns for various aspects of animal body plans: type of symmetry, presence of pseudocoelom or coelom, presence of tube-within-a-tube plan, segmentation, etc. Fill in each of these character columns for each phylum based on the observations you have made today.

IMPORTANT TERMS

annelid	crop	gonad	posterior
anterior	dorsal	intestine	pseudocoelom
anus	embryology	jellyfish	radial symmetry
bilaterally symmetrical	epidermis	medusa	roundworm
blastula	esophagus	mollusk	schistosome
brain	eyespot	mouth	segment
cilia	flatworm	nerve cord	segmented worm
cnidarian	fluke	parapodium	tapeworm
cnidocyte	foot	pharynx	tentacle
coelom	gastrovascular cavity	planaria	trochophore
coral	gill	planula	ventral
coral reef	gizzard	polyp	

Laboratory Report The Rise of the Animal Kingdom

FACTUAL CONTENT

Define the following terms. Brief definitions are given in the chapter, but you may also wish to consult your lecture notes and text for a better idea of how the terms are used. Pay close attention to those words that look similar but have different meanings – you will want to be able to tell them apart if they appear next to each other on a multiple-choice exam.

annelid

anterior

anus

bilaterally symmetrical

blastula

brain

cilia

cnidarian

cnidocyte

coelom

coral

coral reef

crop

dorsal

embryology

epidermis

esophagus

eyespot

flatworm

fluke

foot

gastrovascular cavity

gill

gizzard

gonad

intestine

jellyfish

medusa

mollusk

mouth

nerve cord

parapodium

pharynx

planaria

planula

polyp

posterior

pseudocoelom

radial symmetry

roundworm

schistosome

segment

segmented worm

tapeworm

tentacle

trochophore

ventral

In addition to knowing what these terms mean, you should be able to identify, on both diagrams and prepared specimens, those that are anatomical structures.

After the common name of each group of animal, write the formal scientific name of the corresponding phylum:

corals and jellyfish

flatworms

mollusks

roundworms

segmented worms

Label the anterior, posterior, ventral, and dorsal sides of the animal:

EXERCISES - Record your data and answer the questions below.

A. Cnidarian structure and behavior

1. **Hydra**: Sketch the appearance of a living hydra:

   Do they show evidence of very coordinated movement?

2. **Cnidarian colonies**: Sketch an *Obelia* colony:

   How does the colony compare to a single hydra?

   Can you distinguish the feeding polyps from those specialized for producing medusae? Label them on your sketch.

3. **Cnidarian medusae**: Sketch an *Obelia* medusa:

How does it compare to a polyp?

4. **Planula larvae**: Sketch a cnidarian planula:

B. Flatworm structure and behavior

1. **Planarian morphology**: Sketch a prepared slide of a planaria and label the gastrovascular cavity and pharynx:

How does the gastrovascular cavity compare to that of a hydra?

2.   **Planarian behavior**: Place a living planaria in a deep well slide or shallow dish. Do you see a head with eyespots?

When it moves, what direction does it go?

How does it respond to light?

C. Roundworm dissection:

Sketch the anatomy of an *Ascaris* worm and label the digestive system:

D. Earthworm dissection

1.   Examine the external features of a preserved earthworm – can you see the mouth and anus?

How might the bristles help the worm move?

2. Sketch the internal anatomy of the earthworm. Label the major sections of the digestive system, the dorsal blood vessel, and the hearts. Label the ventral nerve cord and brain if you found them:

E. Parapodia

1. Sketch or describe the variety of preserved parapodia on display:

2. Sketch the parapodia from a prepared slide, labeling the chitinous bristles:

F. Mollusk diversity

1. **Snail behavior**: Sketch or describe the snails in an aquarium tank:

Label the various body regions. Can you see the head portion of the foot? Where is the mouth?

2. **Mollusk diversity**: Sketch or describe the various preserved mollusks on display:

Label the shell (if present) and the foot and body regions of the basic mollusk body plan in each specimen. How are these modified in each animal?

G. Body plans

Construct a table to compare body plans:

phylum	symmetry	pseudocoelom/ coelom	tube-within -a-tube plan	segmentation

Fungi, Vascular Plants, and the Colonization of Dry Land

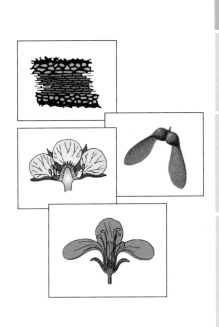

20

OBJECTIVES

New Skills

In performing the exercises in this chapter, you will learn how to dissect flowers in order to reveal their internal anatomy.

Factual Content

In this chapter you will learn
1. some terms about symbiosis in general and symbiotic fungi in particular.
2. the names of two types of vascular tissue found in plants.
3. the names and characteristics of six important divisions of vascular plants.

New Concepts

In doing the exercises in this chapter, you will
1. observe how fungi interact with algae and plants to modify terrestrial environments.
2. observe how specialized vascular tissues allow sporophytes to attain large sizes by transporting materials and providing physical support.
3. continue to study alternation of generations.
4. observe how plant evolution has involved wrapping reproductive structures in more and more layers of protective tissues.

INTRODUCTION

For 90 percent of the earth's history, the continents were barren places. Fossils from that era show us nothing but aquatic organisms – there were no creatures living on dry land. Without plants, there would have been little organic material entering the soil, and everywhere the landscape must have looked like a desert of rocks, sand, and clay. Doubtless there were areas where bacteria made the most of puddles and other moisture; even today we find living cultures buried in ancient rocks where water is scarce. But these could do little more than slowly pile up mats of scummy material.

This began to change around four to five hundred million years ago when the first plants appeared. Most people think of plants as "primitive" compared to animals, but all of the major animal phyla had already evolved in the oceans before any plants existed. Thus, the plant kingdom is the youngest and most "modern" of all of the kingdoms of life.

THE IMPORTANCE OF FUNGI

When the first plants evolved from charophytes and moved onto land, they were not alone. They were accompanied by fungi, with whom they forged a critical partnership. To this day most plants will not grow very well unless they have fungi intermingled with their roots.

We first encountered fungi when we considered saprotrophy. At that time we focused on the role of fungi in decaying and recycling organic materials in the environment. Fungi have other ecological roles as well. In many cases fungi form symbiotic relationships with other creatures. Symbioses can work in two ways: the relationship can be harmful to one partner (i.e., parasitism) or both part-

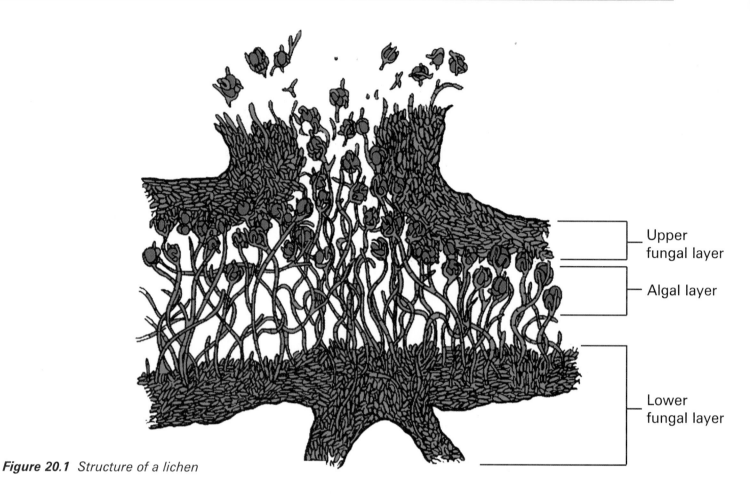

Figure 20.1 *Structure of a lichen*

ners can benefit, a kind of symbiosis called **mutualism**.* Fungi form both types of relationships; many fungi are disease-causing parasites, while others form mutualistic partnerships called **lichens** and **mycorrhizae**.

As you saw in chapter 16, the body of a fungus looks like a mass of thin threads. Such a fungal body is called a **mycelium** (many words relating to fungi are based on the Greek word for mushroom, *mykes*). A lichen is a fungus with photosynthetic algae living amongst the threads of its mycelium (see figure 20.1). The mycelium protects the algae and keeps them moist, while the fungal exoenzymes free up nutrients from the environment that both partners can use. The algae provide the fungus with carbohydrates from photosynthesis. Lichens are very self-sufficient creatures and tend to be among the first organisms to colonize inhospitable environments. They are very important in arctic and mountain environments, where they break down rock and build soil. They may have played a similar role in making land ready for plants hundreds of millions of years ago.

* A third type of symbiosis is at least theoretically possible, in which one partner benefits and the other is neither harmed nor helped. This is labeled *commensalism*. In practice, it can be difficult to determine whether a symbiont is being affected slightly or not at all, so commensal associations are hard to prove. For example, mycologists have long debated whether lichens are mutualistic or commensal associations.

Fungi that live in soil often infect underground plant parts (usually roots, but sometimes underground stems or other parts). These fungi are non-parasitic and do not harm the plants – in fact, they form mutualistic associations called mycorrhizae (*rhiza* is Greek for root). The mycorrhizal mycelium spreads extensively underground, nourished partly by carbohydrates produced by the plant but also saprotrophically on organic matter in the soil. The fungus can also channel nutrients (especially minerals like phosphate) to the infected plants, and plants with mycorrhizae tend to grow better than uninfected plants. About 80 percent of all plants normally participate in mycorrhizal associations. The availability of fungal partners can limit the habitats where plants can thrive.

Different types of fungi infect plants in different ways. Figure 20.2 illustrates how some form a thick covering around the outside of a root, while others can actually penetrate individual root cells. The bulk of a mycorrhizal mycelium is hidden underground, but some fungi produce reproductive structures such as mushrooms that may rise above ground for short periods of time. The rings of mushrooms often found surrounding trees are signs that the tree's roots have mycorrhizae. The earliest fossil impressions of plant roots have mycorrhizae, suggesting that they were important in helping primitive plants survive in ancient soils.

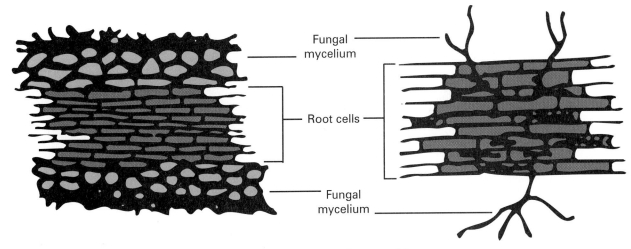

Figure 20.2 *Types of mycorrhizae*

VASCULAR TISSUE

With a waxy cuticle and appropriate modifications to their reproductive systems, plants can live on land without drying out. If these are the only adaptations the plants have, they are limited to relatively small sizes and damp habitats. This is the case with mosses and hornworts. All other plants have one more adaptation – **vascular tissue**.

Vascular tissue allows a plant to transport materials throughout its body, including water entering the plant through the roots and carbohydrates produced by photosynthesis in the leaves. Vascular tissue also forms the woody skeleton that supports large plants.

Vascular tissue or no, gametophytes must be small so that sperm can have a reasonable chance of swimming to an egg. Therefore gametophytes have no real use for vascularization. Vascular tissue occurs only in sporophytes, which are thus able to grow to tremendous heights. Due to alternation of generations, the spores of vascular sporophytes do not germinate into other sporophytes but rather into tiny gametophytes. The gametophytes of vascular plants are smaller than mosses or hornworts, often even microscopic, and are not easy to find. They reproduce sexually and die giving birth to a new generation of sporophytes.

In animals the vascular system is often called the circulatory system, because it circulates blood throughout the animal. Plants are different – their vascular systems are not closed and do not circulate. Instead they have two separate one-way conduction systems. **Xylem** carries water up from the roots to the leaves where it constantly evaporates into the air. **Phloem** brings carbohydrates down from the photosynthetic cells in the leaves and stem to the roots, where it is consumed. Because this system is open and continuously losing moisture, plants generally need a regular

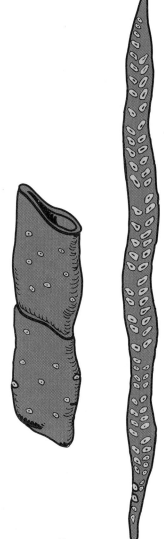

Figure 20.3 *Xylem cells*

223

Courtesy of Champion International Corporation.

Figure 20.4 *Diagram of the layers of tissue in the trunk of a conifer*
  1. outer bark
  2. phloem
  3. growth region (produces new xylem and phloem)
  4. xylem (wood)

source of water to survive. Both xylem and phloem are complex tissues composed of several types of specialized cells, but in each case the functional cells are long and thin like pipes (figure 20.3). Xylem cells are dead and hollow at maturity and literally function like pipes. Wood is made up of layers of old xylem (xylem comes from the Greek word *xylon*, which means wood). On the other hand, phloem cells must be alive to function. Phloem does not build up in thick layers but is replaced each year with fresh tissue. In trees the living phloem is outside the woody core just under the bark (figure 20.4). As long as the phloem is alive the tree can survive, even if most of the old xylem rots away, leaving a hollow tree. However, if the bark is damaged and the thin layer of phloem dies, the tree will die no matter how much wood remains.

All true stems, roots, and leaves are serviced by threads of xylem and phloem. Young plants are **herbaceous**, or soft and green. If they are **annuals** – plants that only survive a single growing season – they will not accumulate enough xylem to become **woody**. **Perennials** live for several years (some live for centuries or even millennia!) and often become woody. Some perennials, like palms, get large and fibrous but nevertheless are unable to produce wood because their xylem tissue is scattered throughout their stems in bundles instead of continuous rings.

SEEDLESS VASCULAR PLANTS
Kingdom Plantae, Divisions Pterophyta and Lycophyta

Most vascular plants reproduce by seeds. Seeds are a complex variation on alternation of generations and will be described later. The first vascular plants did not make seeds. These plants have many modern descendants, of which the most common are **ferns** (Division Pterophyta).

The life cycle of a typical fern is illustrated in figure 20.5. Fern gametophytes are tiny (almost microscopic) and heart-shaped. They grow on top of the ground and are photosynthetic, but they may also have mycorrhizal fungi reaching into their bodies. In most species the gametophyte bears both antheridia and archegonia. When a sperm swims to an archegonium and fertilizes an egg, the zygote begins to grow into a new sporophyte.

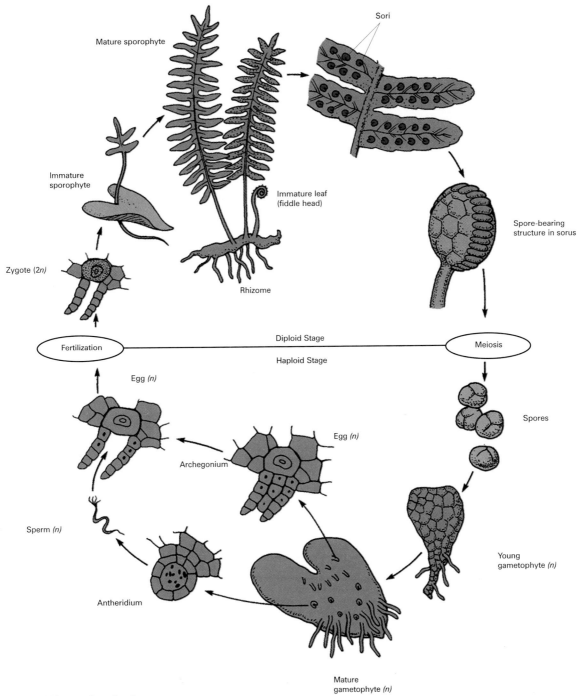

Figure 20.5 *Life cycle of a fern*

The sporophyte will be bigger than the gametophyte, so the gametophyte will die as the sporophyte grows. The young sporophyte develops into a horizontal stem called a **rhizome** and begins to sprout leaves. Most ferns bear small umbrella-shaped structures called **sori** on the undersides of their leaves. Beneath the sori are tiny bulbous structures full of haploid spores. These are released to the wind, and where they land and germinate, the life cycle begins again with new gametophytes. Although fern gametophytes tend to look alike, fern sporophytes are quite diverse, ranging from tiny aquatic plants less than a centimeter long to 100-foot tall tropical tree ferns.

Lycopods (Division Lycophyta) are similar to ferns. Some species are frequently found growing on the floor of coniferous forests and are often referred to as "ground pines" (figure 20.6).

Figure 20.6 *A club moss,* Lycopodium

Mature lycopod sporophytes bear densely leaved branches called **strobili**. Each leaf of the strobilus is a vessel full of spores (figure 20.7). The entire strobilus looks like a small, soft club, suggesting the common name for lycopods, "**club mosses**" (they are, of course, not true mosses). The spores germinate into small gametophytes that look like pale lumps of tissue. These gametophytes live underground and are totally dependent on mycorrhizae for nourishment. In this case the plants are parasites of the fungi!

Although they are not very numerous today, lycopods and ferns were once quite common. Coal was formed from decaying forests of huge tree lycopods (tree lycopods are now extinct) and ferns. Three hundred million years ago, lycopod and fern forests were the dominant terrestrial environments, the landscapes into which ventured the first terrestrial animals. Those forests were also the place where a new innovation in the plant kingdom occurred: seeds.

SEEDS

The best way to understand seeds is to compare them to ferns and lycopods, which do not make seeds. The gametophytes of ferns and lycopods are tiny – the prominent form of these plants is the sporophyte. And of course the sporophyte reproduces via spores. Is the seed, then, a kind of spore? Not exactly.

So far the sporophytes we have looked at have produced a single type of spore. In some plants, however, the sporophyte produces two types of spores: large **megaspores** and small **microspores**. Megaspores always grow into female gametophytes and microspores always grow into male gametophytes. Seed plants produce megaspores and microspores. But these spores are not released as they would be by ferns and lycopods. Instead they remain trapped inside specialized leaves and grow into microscopic gametophytes inside the sporophytes.

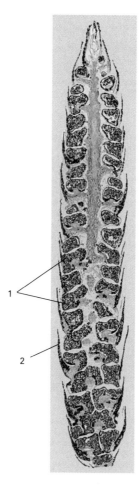

Figure 20.7 *Longitudinal section of a* Lycopodium *strobilus*
1. spores
2. spore-bearing leaves

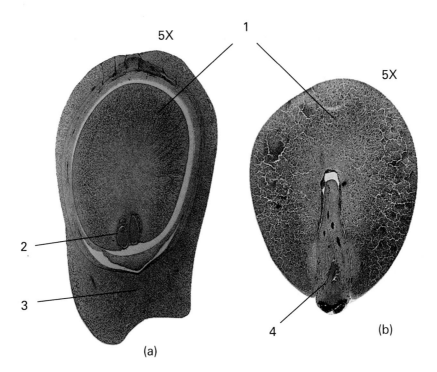

Figure 20.8 *(a) An ovule from the cycad* Zamia *(b) Developing* Zamia *seed, with seed coat (integuments) removed*
   1. female gametophyte
   2. archegonium
   3. integument
   4. sporophyte embryo

The megaspore is trapped by a layer of leaves called an **integument**. Therefore the female gametophyte develops inside the integument. A female gametophyte surrounded by an integument is called an **ovule**. Fertilized ovules turn into seeds (figure 20.8*a*).

Meanwhile, the microspore grows into a microscopic gametophyte called **pollen**. When the pollen is mature, the sporophyte finally lets go of it, and it is free to be carried toward an ovule (either by wind or animals). The pollen grows toward a minute opening in the integument of the ovule until it reaches the archegonium of the female gametophyte. At this point the pollen can release its sperm into the archegonium to fertilize the egg. The embryo of a new sporophyte then begins to grow inside the gametophyte. An ovule that contains an embryo is a **seed** (figure 20.8*b*). The seed is actually an entire family in one small package: parent (sporophyte integument), daughter (female gametophyte), and grandchild (new sporophyte embryo).

With seeds we have come to the exact opposite of mosses. In the moss life cycle the dominant plant is the gametophyte, and the sporophyte is parasitic on the gametophyte. In seed plants the visible plant is the sporophyte, and the gametophytes are parasites on them.

The oldest fossil seeds are from about 350 million years ago. At that time the great coal forests of the Carboniferous Period were full of spore-bearing plants. Seed plants gradually became more and more common, so that by the time dinosaurs appeared, they had largely replaced the ferns and lycopods.

There are several types of seed plants. **Cycads** (Division Cycadophyta) were very common during the age of dinosaurs but are rare now, living exclusively in tropical and subtropical environments. Cycads are large plants and produce big seeds (the specimens in figure 20.8 are from a cycad). The Chinese **ginkgo**, an unusual tree whose nearest relatives have all been extinct for millions of years, is another kind of seed plant. It is the only surviving member of Division Ginkgophyta. The most common seed plants today are **conifers** and flowering plants. Indeed, these plants are so common that we define most of the earth's environments, from grasslands to tropical forests, by their presence.

CONIFERS
Kingdom Plantae, Division Coniferophyta

Most divisions of seed plants are called **gymnosperms** because they do not make their seeds inside a flower (*gymnosperm* comes from the Greek *gymnos*, naked, and *sperma*, seed). Cycads and ginkgoes are gymnosperms, but they are not as familiar as conifers. Conifers live in many different environments and are the only gymnosperms that form large forests. The better-known conifers include pines, spruces, yews, redwoods, and junipers. Although the word *conifer* is often taken to be synonymous with "evergreen," this is an oversimplification: some, such as the larch and bald cypress, are **deciduous** and lose their leaves every fall.

As their name implies, most conifers bear ovules and pollen in **cones**. A cone is essentially the same thing as a strobilus, although

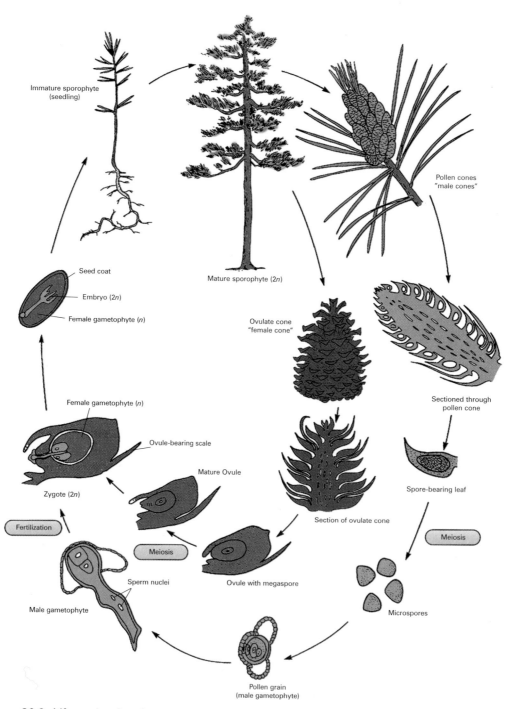

Figure 20.9 *Life cycle of a pine*

it is larger and its leaves are tough and fibrous. The similarity to strobili is best seen in pollen cones (figure 20.9). Ovulate cones are larger than pollen cones and persist on the tree for two to three years while the gametophytes grow and produce seeds. Conifer seeds are smaller than cycad seeds and often have small wings on them. These attributes help the seeds to disperse over a wide area.

ANGIOSPERMS
Kingdom Plantae, Division Anthophyta

You have doubtless seen various conifers in your life, and probably ferns and mosses, too. Yet most of the plants you know are flowering plants. Flowering plants include everything from grasses to crop plants to towering oak trees. True, not all of these produce showy blossoms like one sees in a typical flower garden. They do, however, make seeds inside special arrangements of modified leaves, and these organs, no matter what size or color they are, are **flowers**.

So far the story of plant evolution has been one of surrounding reproductive structures in new layers of tissue. In the charo-

phytes, sterile cells surrounded the eggs; these became the archegonia in land plants. In gymnosperms, an integument surrounded the female gametophyte to form an ovule. Flowers continue this trend. At the center of a flower are one or more leaves that wrap around the ovules. These protective leaves are **carpels**. The bases of the carpels join together to form a vessel – the **ovary** – which hold the ovules and the seeds they become. For this reason, flowering plants are called **angiosperms** (from the Greek words *angion*, vessel, and *sperma*, seed). The tips of the fused carpels extend as a long **style** ending in a **stigma**, which is specialized for collecting pollen (figure 20.10). Stigma, style, and ovary together constitute the **pistil**. The simplest pistil is a single carpel curled into a tube; some flowers have pistils made of several carpels linked together in a continuous structure, something like the staves of a barrel.

Surrounding the pistil are the **stamens**. A stamen consists of an **anther** supported by a thin **filament**. The anthers are where the microspores form and turn into pollen. The stamens are surrounded by a ring of leaves, often brightly colored, called **petals**. Around the petals is a second ring of leaves, the **sepals**. A flower that has all of these parts is called **perfect**. **Imperfect** flowers

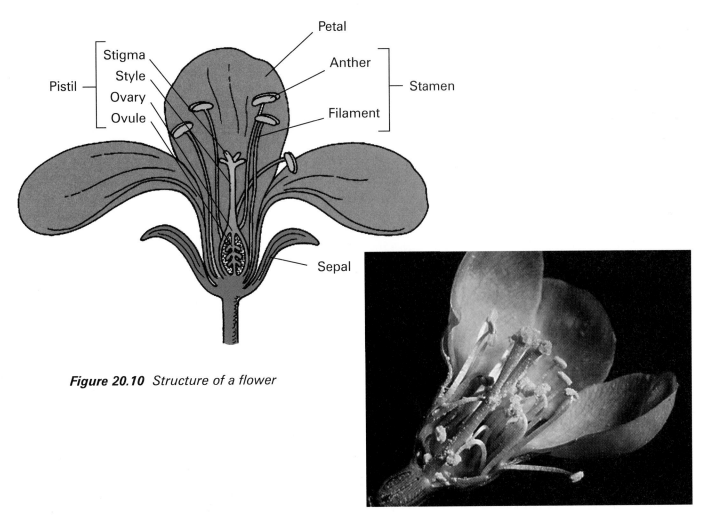

Figure 20.10 *Structure of a flower*

Figure 20.11 *A dissected rose*

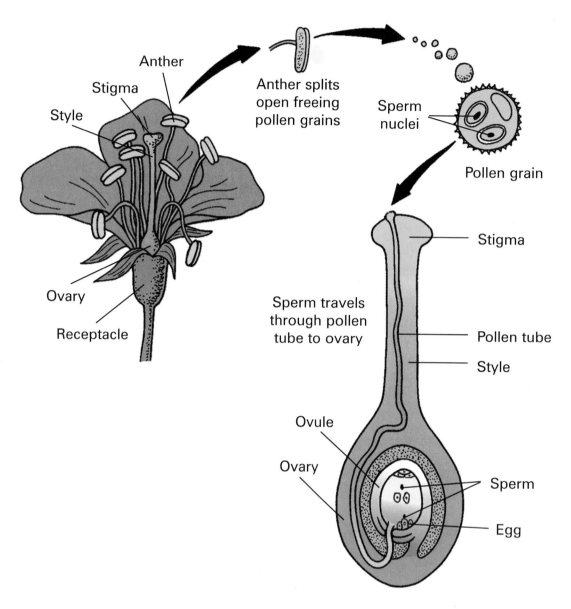

20.12 *Diagram illustrating pollination in angiosperms*

lack one or more floral parts. For example, some plants have pistils and stamens in separate flowers. The pistillate flowers lack stamens and the staminate flowers lack pistils. Other kinds of imperfect flowers may be missing petals or sepals. Figure 20.11 shows how the parts of a perfect flower appear in an actual specimen.

Many angiosperms bear their flowers singly, as does the familiar rose. Other species have their flowers in clusters. A floral cluster is called an **inflorescence**. The inflorescences of asters, daisies, sunflowers, and dandelions are so densely packed that they look like a single blossom, but close inspection reveals that each apparent "petal" is actually a small tubular flower with its own pistil inside.

The oldest fossil flowers are roughly 100 million years old. They appeared during the age of dinosaurs, when conifers and cycads had already replaced the ancient forests of lycopods and ferns. Angiosperms were very successful and slowly began to replace many gymnosperm species. They diversified greatly after the dinosaurs became extinct and have been the earth's major form of plant life for the past 70 million years.

The structure of a flower gives angiosperms several advantages over gymnosperms. For one thing, the leaves around the stamens – the petals and sepals – can be specialized to attract insects, birds, or bats with bright colors or secreted nectar, enticing the animals

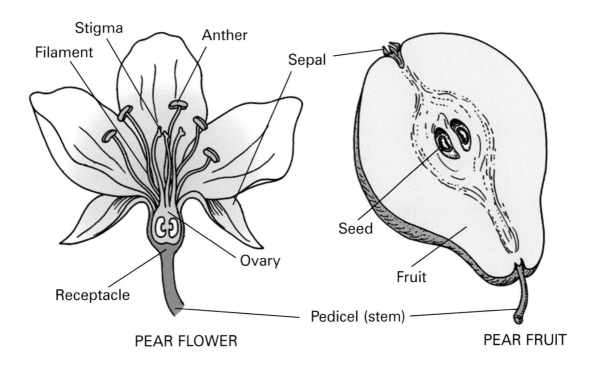

PEAR FLOWER

PEAR FRUIT

Figure 20.13 *The flower and fruit of a pear. The fruit develops from several flower tissues.*

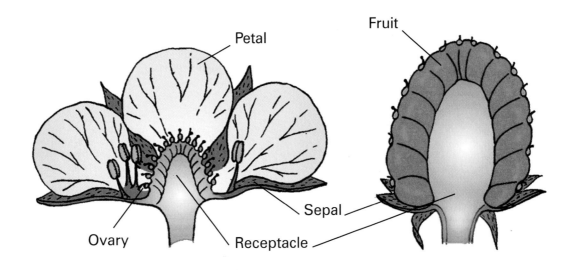

Figure 20.14 *The flower and fruit of a strawberry. The strawberry is an aggregate fruit.*

to carry pollen from one flower to another. This tends to be more efficient than wind pollination.

Another advantage of flowers arises from what happens to the carpel after the ovules are fertilized and begin to turn into seeds. Figure 20.12 shows how a pollen grain must grow a long tube all the way through the pistil in order to deliver sperm to the egg within the ovule. The ovules remain in the ovary after fertilization. As the seeds develop, the carpels of the ovary change too. They swell into an enlarged seed-bearing organ called a **fruit**. A fruit is a mature ovary containing seeds (figure 20.13).

By this definition, many common "vegetables" are fruits: tomatoes, green peppers, cucumbers, pumpkins, squash, green beans,

etc. Nuts and acorns are also fruits. As fruits they all start out as flowers.

Different kinds of fruits develop in different ways. **Simple fruits** — for example, nuts, grains, legumes, tomatoes, and grapes — are derived from a single ovary. The outer portion of a simple fruit, whether hard and dry or soft and fleshy, is made up of ovary tissue. An **accessory fruit** contains tissue from additional floral parts. Apples and pears (figure 20.13) are both accessory fruits. The bulk of these fruits are derived from the stem receptacle that holds the ovary; only the central portion surrounding the seeds comes from the ovary itself. **Aggregate fruits** develop from a single flower with many pistils. Raspberries, blackberries, and strawberries (figure 20.14) are all aggregate fruits. Finally, **multiple fruits** are derived from several flowers that grow together.

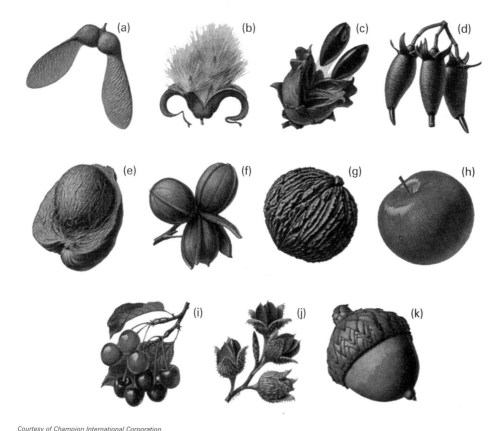

Courtesy of Champion International Corporation.

(a) Maple: The winged fruits of a maple fall with a spinning motion that may carry it hundreds of yards from the parent tree.
(b) Willow: The air-borne fruits of a willow may be dispersed over long distances.
(c) Witch Hazel: Mature seeds of the witch hazel tree are dispersed up to 10 feet by forceful discharge.
(d) Mangrove: The fruits of this tropical tree begin to germinate while still on the branch, forming pointed roots. When the seeds drop from the tree, they may float to a muddy area where the roots take hold.
(e) Coconut: The buoyant, fibrous husk of a coconut permits dispersal from one island or land mass to another by ocean currents.
(f) Pecan: The husk of a pecan provides buoyancy and protection as it is dispersed by water.
(g) Black walnut: The encapsulated seed of the black walnut is dispersed through burial by a squirrel or floating in a stream.
(h) Apple: The seeds of an apple tree may be dispersed by animals that ingest the fruit and pass the undigested seeds hours later in their feces.
(i) Cherry: Moderate sized birds, such as robins, may carry a ripe cherry to an eating site where the juicy pulp is eaten and the hard seed is discarded.
(j) Beech: Seeds from a beech tree are dispersed by mammals as the spiny fruits adhere to their hair. In addition, many mammals ingest these seeds and disperse them in their feces.
(k) Oak: An oak fruit may be dispersed through burial by a squirrel.

Figure 20.15 *Examples of seed dispersal*

The best known example of this is the pineapple – the bumps on its surface are the vestiges of the many flowers.

Fruits can be important aids to seed dispersal (a few examples are illustrated in figure 20.15). Some fruits have wings or light, fluffy fibers that can keep them suspended in the the wind for long distances. Others have hooks and barbs that cling to the fur or feathers of animals, allowing the seeds to "hitchhike" to new habitats. Still other fruits become fleshy and are eaten by animals who then spread the seeds in their feces. Some seeds are so specialized that they cannot germinate until they have passed through the digestive tract of a particular animal! The mobility fruits provide to seeds is a major reason that angiosperms have been able to colonize so many different environments.

EXERCISES WITH FUNGI AND VASCULAR PLANTS

A. Lichens and mycorrhizae

1. Observe some of the lichens on display. Are any of them growing on rocks? Do they have an obvious source of water? How does their form compare to the mycelia you saw growing on moist agar in chapter 16? How many different colors are present in the specimens?

2. Examine a prepared slide of a root infected with mycorrhizal fungi. Can you identify the mycelium as distinct from the plant cell walls? Usually they are stained different colors. What chemical differences might account for these different reactions to stains? Does the mycelium in your specimen extend into the root cells or is it limited to the outside of the root?

B. Vascular tissue

1. Examine a prepared slide of macerated wood. This is xylem that has been treated so that the individual cells separate from each other. Observe the thin tubular shape of the cells, and look for pores in the cell walls that allow water to pass from cell to cell.

2. Next observe a slide of a cross section of a woody stem. Identify the rings of xylem. How many years of growth can you observe? Find the layer of phloem tissue outside the xylem. Is it as thick as the xylem?

C. Fern life cycle

1. Survey the various fern sporophytes on display and compare their shapes. Look for sori.

2. Look at the prepared slides showing fern gametophytes. Do they contain mycorrhizae? How do the fern gametophytes compare to the fern sporophytes? How do the fern gametophytes compare to moss gametophytes? Identify the archegonia and antheridia and observe how the young sporophyte grows out of the archegonium.

D. Lycopod structures

1. Study living or preserved *Lycopodium* and compare the body to the fern sporophytes. Identify the strobili and study one under the microscope.

2. Examine the preserved lycopod gametophyte. Does it look at all like the sporophyte?

E. Cycad structures

Cycads are relatively rare today, but some species of *Zamia* are sometimes grown in greenhouses as ornamental plants. Their ovules are exceptionally large and are therefore good specimens for study.

1. If a living cycad is available, observe its structure. Examine the large compound leaves. How do they compare to fern leaves? Look for any cones. How do they compare to pine cones?

2. Examine a prepared slide of a cycad ovule. Identify the gametophyte in the center and observe how it is surrounded by integuments. Diagram this in your notes.

F. Conifer cones

1. Compare the small pollen cones to the larger ovulate cones. Pollen cones are usually small, and they frequently appear in dense clusters. Ovulate cones are larger than their pollinate counterparts. (This is relative: in some species the ovulate cones can be pretty small, but the pollen cones will still be smaller.) Ovulate cones often have sticky resin on them to trap wind-blown pollen.

2. Conifer ovules can be seen on prepared slides of ovulate cones. Note how the ovules rest on top of the cone scales. Can you find the integuments as easily as you did in exercise E?

3. Look at a slide of a pollen cone. Note how the leaves are expanded like balloons to hold the pollen. How does this compare with the lycopod strobilus?

G. Angiosperm structures

1. You will be given a flower to dissect. First examine the whole flower without a microscope. Is it symmetrical or asymmetrical? Asymmetrical flowers are often highly modified. Next determine whether your flower is perfect or imperfect. Try to identify all of the parts in figure 20.10.

2. Pull off the petals on one side and examine the flower under the dissecting microscope. Identify the stamens and pistil.

3. Remove the petals and stamens to expose the pistil. If your flower's pistil is large enough, open the ovary and find the ovules.

4. Survey the various fruits on display. Find examples of simple fruits, accessory fruits, aggregate fruits, and multiple fruits.

H. Comparison of sporophytes and gametophytes

1. Divide a page in your report or notebook into two columns. Label one "sporophyte" and the other "gametophyte."

2. Review your notes of observations for chapter 18. Describe or sketch the sporophytes and gametophytes of mosses and hornworts in the appropriate columns of your table.

3. Now continue the table with the sporophytes and gametophytes of ferns, lycopods, conifers, and angiosperms.

4. What trends or generalizations can you derive from this comparison?

IMPORTANT TERMS

accessory fruit	filament	microspore	pollen
aggregate fruit	flower	multiple fruit	rhizome
angiosperm	fruit	mutualism	seed
annual	ginkgo	mycelium	sepal
anther	gymnosperm	mycorrhizae	simple fruit
carpel	herbaceous	ovary	sori
club moss	imperfect	ovule	stamen
cone	inflorescence	perennial	stigma
conifer	integument	perfect	strobilus
cycad	lichen	petal	style
deciduous	lycopod	phloem	vascular tissue
fern	megaspore	pistil	woody
			xylem

Laboratory Report Fungi, Vascular Plants. and the Colonization of Dry Land

FACTUAL CONTENT

Define the following terms. Brief definitions are given in the chapter, but you may also wish to consult your lecture notes and text for a better idea of how the terms are used. Pay close attention to those words that look similar but have different meanings – you will want to be able to tell them apart if they appear next to each other on a multiple-choice exam.

accessory fruit	ginkgo
aggregate fruit	gymnosperm
angiosperm	herbaceous
annual	imperfect
anther	inflorescence
carpel	integument
club moss	lichen
cone	lycopod
conifer	megaspore
cycad	microspore
deciduous	multiple fruit
fern	mutualism
filament	mycelium
flower	mycorrhizae
fruit	ovary

ovule   simple fruit

perennial   sori

perfect   stamen

petal   stigma

phloem   strobilus

pistil   style

pollen   vascular tissue

rhizome   woody

seed   xylem

sepal

In addition to knowing what these terms mean, you should be able to identify, on both diagrams and prepared specimens, those that are anatomical structures.

After the common name of each group of plant, write the formal scientific name of the corresponding division:

conifers

cycads

ferns

flowering plants

ginkgoes

lycopods

EXERCISES - Record your data and answer the questions below.

A. Lichens and mycorrhizae

1.   Sketch or describe the lichens on display:

Are any of them growing on rocks?

Do they have an obvious source of water?

How does their form compare to the mycelia you saw growing on moist agar in chapter 16?

How many different colors are present in the specimens?

2.   Sketch a prepared slide of a root infected with mycorrhizal fungi, labeling the mycelium and the plant cell walls:

What chemical differences might account for the different reactions to stains?

Does the mycelium in your specimen extend into the root cells or is it limited to the outside of the root?

B. Vascular tissue

1. Sketch a prepared slide of macerated wood:

    Label the pores in the cell walls.

2. Sketch a slide of a cross-section of a woody stem:

    Label the phloem and rings of xylem.

    How many years of growth can you observe?

Is the phloem tissue as thick as the xylem?

C. Fern life cycle

1. Sketch or describe the various fern sporophytes on display and compare their shapes:

Label any sori.

2. Sketch the fern gametophytes:

Label the archegonia, antheridia, and the young sporophyte growing out of the archegonium.

Do any of the gametophytes contain mycorrhizae?

How do the fern gametophytes compare to the fern sporophytes?

How do the fern gametophytes compare to moss gametophytes?

D. Lycopod structures

1. Sketch or describe living or preserved *Lycopodium* and compare the body to the fern sporophytes:

    Sketch a strobilus:

2. Sketch or describe the preserved lycopod gametophyte. Does it look at all like the sporophyte?

E. Cycad structures

1. Sketch or describe a living cycad if available:

How do the cycad leaves compare to fern leaves?

Look for any cones. How do they compare to pine cones?

2.   Sketch a prepared slide of a cycad ovule:

Label the gametophyte and integuments.

F.   Conifer cones

1.   Sketch or describe the pollen cones and ovulate cones on display:

2.   Sketch a prepared slide of an ovulate cone:

Can you find the integuments as easily as you did in exercise E?

3.   Sketch a slide of a pollen cone:

How does this compare with the lycopod strobilus?

G. Angiosperm structures

1.   Sketch the whole flower without a microscope:

Label the parts illustrated in figure 20.10.

Is your flower symmetrical or asymmetrical?

Is your flower is perfect or imperfect?

2.   Sketch or describe the various fruits on display:

H. Comparison of sporophytes and gametophytes

Complete the table with sketches or descriptions of sporophytes and gametophytes:

plant	sporophyte	gametophyte
mosses		
hornworts		
ferns		
lycopods		
conifers		
angiosperms		

What trends or generalizations can you derive from this comparison?

The Origins of Land Animals

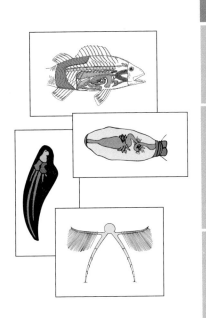

21

OBJECTIVES

New Skills

No new skills are required for these exercises.

Factual Content

In this chapter you will learn
1. some more terms about embryology.
2. the names and characteristics of two phyla of animals whose members live on land.
3. the names of some important structures found in representative specimens of these animal phyla.

New Concepts

In doing the exercises in this chapter, you will
1. observe some of the diversity possible for terrestrial animal body plans.
2. compare the segmentation and appendages of arthropods and annelids.
3. observe how insects are arthropods modified for life on dry land.
4. observe the basic body plan of chordates and how it is modified in several types of chordates.
5. observe the basic skeletal pattern found in the limbs of terrestrial vertebrates.

INTRODUCTION

Together, plants and fungi made land habitable. Through photosynthesis, plants produced large amounts of organic material that fungi eventually broke down into rich humus, turning barren sand into fertile soil. As soil improved, the plants and fungi moved farther and farther inland, creating new environments. The seas were already full of a wide variety of animals when this process began. Inevitably, some of them followed the plants and fungi ashore.

The first land animals faced the same problems of desiccation and support that confronted plants when they began to live on land. Some animals, such as earthworms and snails, never acquired adaptations to cope with these difficulties very effectively, and to this day they remain limited to moist environments while most of their relatives remain in aquatic habitats. In this respect, such animals resemble those algae that grow in damp areas but are very sensitive to water availability. Of all the different types of algae, only the charophytes produced descendants that were able to deal with dry environments. Similarly, only two groups of animals have been able to colonize land to a significant extent – the **arthropods** and **vertebrates**.

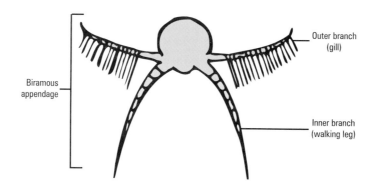

Figure 21.1 *Diagram of a cross section of the extinct arthropod* Marrella

ARTHROPODS
Kingdom Animalia, Phylum Arthropoda

Arthropods are long, segmented creatures similar to segmented worms. They differ from annelids in two important ways. First of all, their bodies are encased in an **exoskeleton** made of chitin. As we saw in chapter 19, segmented worms often have chitinous bristles on their segments, so the material itself is not new. The exoskeleton is a protective covering that surrounds the animal's entire body. Because it seals moisture inside an arthropod's body, the exoskeleton was the critical adaptation that allowed arthropods to move onto land. (The exoskeleton is often called a **cuticle**, just like the waxy covering of plants, which reflects their similar functions.) Many portions of the exoskeleton are soft and flexible for movement, but generally the exoskeleton is thick and rigid. This one structure combats both desiccation and the problem of supporting a body that no longer has the luxury of floating in water.

The second major difference between arthropods and segmented worms is that each segment on an arthropod has a pair of jointed appendages (the word *arthropod* is derived from the Greek *arthron*, joint, and *podos*, foot). In creatures like centipedes, the segments and their appendages are fairly repetitive, but in most arthropods individual segments and their appendages can be highly modified, creating a complex set of specialized tools. Arthropod appendages recall the fleshy parapodia seen on many segmented worms. In many fossil arthropods the appendages are double-branched, or **biramous**. Figure 21.1 illustrates a cross section of the extinct animal *Marrella*. Here each appendage has an outer branch that is fleshy and feathery and probably functioned as a gill. The inner branch points down and is clearly a walking leg. Thus the arrangement of functional sub-parts of a biramous appendage is similar to that found in many parapodia (compare with figure 19.7). Some zoologists see this as a possible evolutionary link, but the biramous appendages in various arthropods (such as extinct trilobites or modern crustaceans) are

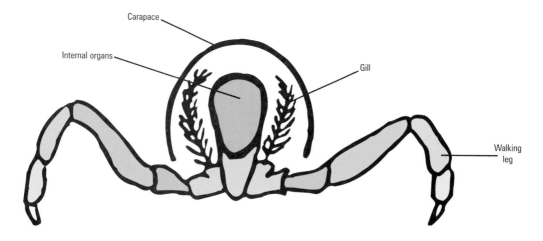

Figure 21.2 *Diagram of a cross section of a crayfish*

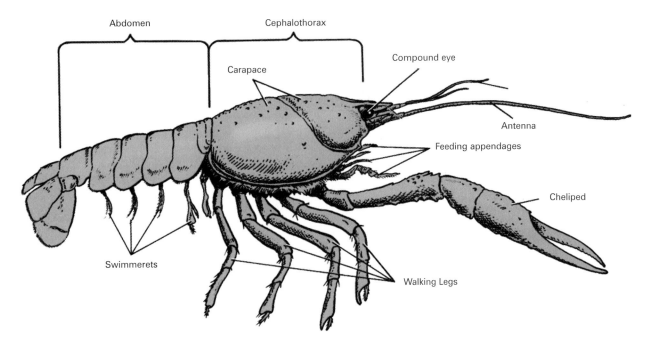

Figure 21.3 *Diagram of a crayfish*

so diverse that the exact relationships between parapodia and biramous appendages remain obscure. Indeed, the evolutionary relationships between the various types of arthropods are quite uncertain, and many zoologists believe that the different types of arthropods evolved independently from wormlike ancestors.

Marrella and the trilobites were aquatic creatures that lived over 500 million years ago. The segmentation in these animals was obvious, and each segment was similar to the next. Other fossil arthropods show the beginnings of specialization in segments and appendages. In the bodies of modern arthropods, segments are often arranged and even fused into specialized regions. The bodies of some arthropods (for example, spiders) show so much fusion that segmentation can be difficult to see. Since the basic arthropod body plan has a pair of appendages on each segment, it is sometimes possible to infer how many segments are fused into a single body region by counting the appendages the region has. The appendages themselves can be specialized for many tasks: fangs, walking legs, stingers, sensory **antennae** ("feelers"), and jaw-like **mandibles** are all common. However, specialization can also cause segments to lose appendages, and this further complicates the study of these creatures.

The arthropod body plan has proved to be very adaptable. In terms of numbers and diversity, arthropods are by far and away the most successful terrestrial animals on earth and are found nearly everywhere.

CRUSTACEANS

Crustaceans constitute a large group of arthropods that remains mostly aquatic. Lobsters, crayfish, shrimps, crabs, and barnacles are all crustaceans. Numerous species of crustaceans are microscopic. Sessile animals like clams and tube worms, which strain small bits of food from water that flows past their mouths, eat large quantities of these tiny crustaceans. The habit of straining food from water is called **filter feeding**.

Crustaceans are characterized by biramous appendages. Figure 21.2 illustrates a cross section of a crayfish body so that the gill branch of the appendage can be seen. Normally the gills of a crayfish are hidden under the **carapace**, a large section of the exoskeleton.

The segments of crustaceans such as crayfish, shrimp, and lobsters are arranged in two body regions: the **cephalothorax** and the **abdomen** (see figure 21.3). The cephalothorax typically exhibits a great deal of fusion, but the segmental character of the abdomen is fairly obvious. The first two pairs of appendages on the cephalothorax are antennae, and after that come several pairs of mouthparts, including a pair of mandibles. Beyond the mouthparts, different crustaceans have different appendages to suit their lifestyles. In crayfish, lobsters, and crabs there is a pair of large pincers called **chelipeds** followed by several pairs of long **walking legs**. Most of the abdominal segments of a crayfish have small **swimmerets**, but the appendages on the final segments are flattened into fins.

Internally, crayfish and their relatives are similar to earthworms. The powerful mandibles crush food and push it through the esophagus to the **stomach**, where digestion begins. The stomach passes food into the intestine, which runs through all of the posterior segments just as it does in worms. Dorsal to the digestive tract is a single heart connected to a large blood vessel. These deliver blood to large open spaces throughout the body and also force blood through the gills that are part of the biramous walking legs. A nerve cord runs ventral to the digestive tract through all of the segments and ends in a small brain just in front of the mouth. The brain receives sensory input from the antennae, the multifaceted **compound eyes**, and other sense organs.

INSECTS

The vast majority of crustaceans are aquatic animals, even though their exoskeletons and appendages give them some ability to function on land. Some species take advantage of these structures to live amphibiously in shoreline habitats, and a few are specialized for life in wet soil, in much the same way that snails and earthworms are. Other groups of arthropods are better adapted for dry conditions. These include the **arachnids** (spiders and scorpions) and **insects**. In fact, insects are easily the most successful of terrestrial animals, having diversified into millions of species with countless individuals. Some insect species are aquatic or semiaquatic, but this appears to be a secondary adaptation from terrestrial ancestors. Fossils indicate that the first insects appeared at roughly the same time that plants were establishing themselves on land, about 400 million years ago.

Insects are similar to crustaceans in having mandibles for jaws, antennae as their first appendages (but only a single pair), and compound eyes. Insects have three distinct body regions instead of two: the **head**, **thorax**, and **abdomen**. On the thorax are three pairs of walking legs. As in crustaceans, the abdomen is the most obviously segmented of the body regions. Insect abdomens have very few appendages on their segments, however, save for the occasional stinger or egg-laying tube.

Unlike those of crustaceans, insect appendages are always unbranched. Instead of gills, insects use a unique system of **trachea** to carry oxygen throughout their bodies like microscopic air ducts. Because of this, the insect cuticle can completely cover the body of the animal and seal in moisture. Air can only enter the trachea through valves in the sides of the abdomen called **spiracles**. The spiracles are only opened as necessary to admit fresh air, so the inside of the insect does not dry out.

Many insects also have one or two pairs of **wings** on their thoraxes. Wings are not derived from segmental appendages – they are outgrowths of the exoskeleton. Insects move their wings by bending the portion of the cuticle to which the wings are attached. Winged insects first appeared about 300 million years

ago, in the great carboniferous forests of giant lycopods. Flight opened up many opportunities for insects, and this was the beginning of the first great period of insect diversification. Later, the origin of seed plants, especially the flowering plants, coincided with another great period of insect speciation.

The anatomy of an insect is much like that of a crayfish, except that the digestive tract is slightly modified and has yet another adaptation to preserve water (figure 21.4). The esophagus expands into a **crop** (analogous to the crop of earthworms), before the stomach. Most digestion occurs in the stomach, so it has several small, finger-shaped **gastric ceca** to increase the size of the digestive space. From the stomach food passes into the intestine. Tiny threads called **Malpighian tubules** lead into the intestine

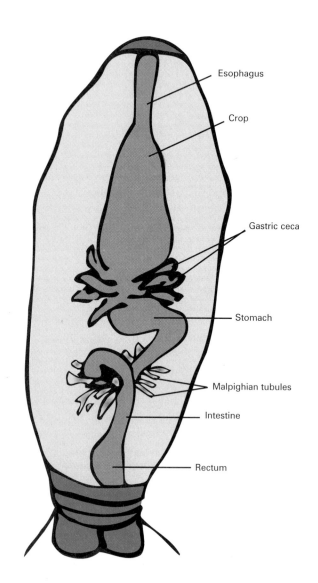

Figure 21.4 *Diagram of the anatomy of a cockroach (dorsal view)*

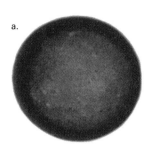
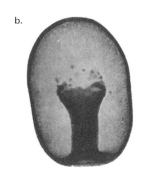
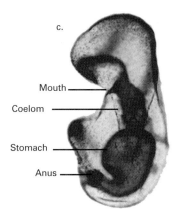

Mouth

Coelom

Stomach

Anus

Figure 21.5 *Embryological stages of a sea star: (a) blastula (b) gastrula (c) early larva*

from the body cavity. These filter waste materials from the body fluids (like a human kidney) and draw them into the intestine so they can be excreted through the anus with any undigested food. Thus insects do not waste water by making urine.

PROTOSTOMES AND DEUTEROSTOMES

So far all of the animals we have looked at are related to some extent by their embryology. Their zygotes begin life by expanding into hollow blastulas (figure 21.5a). At that point cnidarians branch off from the main line, as their blastulas fill with cells to become solid planuloid larvae. In the other groups of animals, one side of the blastula "caves in" to create a cup-shaped structure (figure 21.5b). This embryological stage is called a **gastrula**. The outer layer of the gastrula becomes the skin or epidermis of the animal, and the inner layer becomes the gastrovascular cavity. This is where flatworms part company with the other animals, for their gastrulas develop into adults and never have more than a single opening into the gastrovascular cavity. Animals with one-way digestive tracts – roundworms, segmented worms, mollusks, and arthropods – must go through an additional embryological step: the digestive tract must develop a second opening. In annelids and mollusks this often results in a trochophore larva. Larvae are different in the other groups, but they all end up with a one-way digestive tract. The initial opening formed during gastrulation is the mouth (just as it is in flatworms) and the secondary opening is the anus. Because the mouth forms before the anus, these animals are called **protostomes** (from the Greek *protos*, first, and *stoma*, mouth).

Protostomes share other embryological features, but we need not concern ourselves with these details here. (Strictly speaking, roundworms are not considered true protostomes because they differ in some of these details, which is why they end up with a pseudocoelom instead of a true coelom.) This early development sets several patterns. For example, all protostomes have nerve cords ventral to the digestive tract and hearts or major blood vessels dorsal to the digestive tract. Thus both the embryology and overall body patterns justify treating protostomes as a natural group. And, as the protostomes include the arthropods, with millions of species, and the mollusks, with hundreds of thousands more, the protostomes can truly be considered the main line of evolution in the animal kingdom.

There is a second line of evolution in the animal kingdom, one that exhibits a different pattern of embryology. Animals that develop according to this plan are called **deuterostomes**. Deuterostome zygotes develop into blastulas and gastrulas too, but the way the cells divide is different from the pattern followed by other animals. Moreover, the development of the digestive tract in deuterostomes is upside-down relative to protostomes: the original opening into the gastrovascular cavity becomes the anus and the second opening becomes the mouth (*deuterostome* is derived from the Greek for "second mouth"). The embryos in figure 21.5 are actually from deuterostomes. The photos are aligned the same way so that the larva in figure 21.5c can be aligned with the gastrula from which it arose, showing the relationship between the anus and gastrulation. Protostome and deuterostome embryos also differ in their development after gastrulation, particularly in how the coelom develops. The deuterostome development pattern is so unusual that it seems certain that deuterostomes branched off from the rest of the animal kingdom's line of evolution at a very early stage.

Deuterostomes are definitely not the main line of animal evolution – the number of all deuterostome species put together is less than that of just the mollusks, and it is less than a tenth of the species of arthropods. There are only two important phyla of deuterostomes, the **echinoderms** (sea stars, sand dollars, and their relatives) and **chordates**. However, the chordates include many terrestrial creatures, including humans, and are therefore of special interest to us.

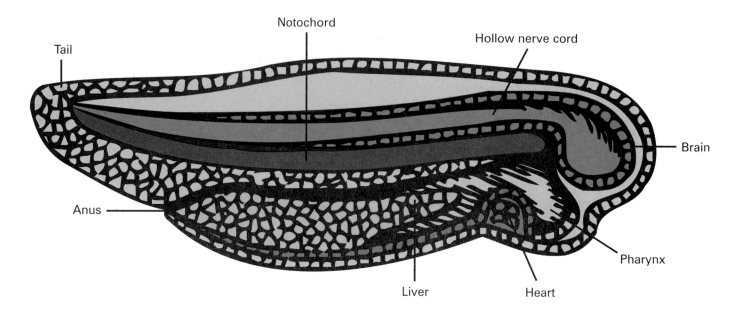

Tail

Notochord

Hollow nerve cord

Brain

Anus

Liver

Heart

Pharynx

Figure 21.6 *Diagram of early tadpole development*

CHORDATES
Kingdom Animalia, Phylum Chordata

All chordates share certain structural features. The name of the group refers to one of these features, the **notochord**. The notochord is linked embryologically to a second chordate characteristic, a **dorsal nerve cord**.

After gastrulation, the epidermis on the dorsal side of a chordate embryo sinks a bit to form a shallow groove. The bits of epidermis on each side of the groove fold up and around, finally forming a hollow **neural tube**. The neural tube is the core of the chordate nervous system, which even in humans is still based on a hollow nerve cord running along the back: the spinal cord. (Notice that the nervous system originates from the same embryological tissue as the skin – not an internal tissue!) The tissue just below the neural tube compacts into a firm rod, the notochord. (Figure 21.6 illustrates how these features are arranged in a developing frog tadpole.) In simple chordates the notochord acts like a skeleton, but in one group of chordates, the vertebrates, it becomes incorporated into a rigid and sturdy skeletal system. Unlike the exoskeleton of arthropods, the internal skeleton of vertebrates is made of living tissue that grows along with the rest of the animal. Arthropods must periodically shed their old cuticles as they grow, and there are physical constraints on the effectiveness of large exoskeletons. The fact that vertebrates have living, internal skeletons explains why they have been able to achieve enormous sizes, while arthropods are generally small.

In chordates the digestive tract does not reach through the entire length of the animal as it does in roundworms, annelids, and arthropods. Instead, a **tail** extends posterior to the anus. All chor-

dates have tails at some point in their lives, although they may be lost as the animal grows.

The body walls of chordate embryos also become segmented like those of annelids and arthropods. Just as in arthropods, chordate segments often fuse into body regions that retain only traces of segmentation. In vertebrates, the rows of bones in the vertebral column and rib cage are vestiges of segmentation. One segmented characteristic that all chordates have is a set of parallel slits in the pharynx. These **pharyngeal slits** are used for filter-feeding in some chordates but have become modified for other uses in vertebrates. Like the tail, the pharyngeal slits of many chordates are lost as the animals mature.

Chordates arose in the seas, just like every other group of animals. Only about 45 percent of modern chordate species live on land, so this group has not embraced the terrestrial lifestyle as completely as the arthropods. One can therefore study many basic chordate characteristics in aquatic specimens that are small and relatively easy to work with.

The oceans are home to the only chordates that are not vertebrates. **Lancelets** are examples of such animals. These are small creatures that look something like a cross between a worm and a tiny fish (figure 21.7). Their adult bodies exhibit all of the fundamental characteristics of chordates: a nerve cord and notochord running down the back, a tail extending beyond the anus, muscles arranged in segmental bands, and a pharynx perforated by numerous parallel slits. Although lancelets are reasonably good swimmers, they are not very active and tend to bury themselves with just their mouths extending above the sand. In this position they filter-feed by drawing water into their mouths and straining

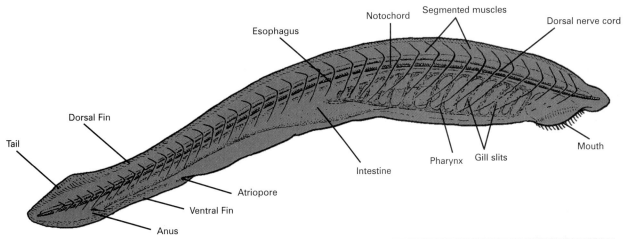

Figure 21.7 *Diagram of the lancelet* Amphioxus

it through their pharyngeal slits. The bits of food caught in the pharynx are swallowed, while the water that was strained through the slits collects in a cavity surrounding the pharynx and is finally expelled through an opening called the atriopore.

FISHES

Vertebrates are animals in which the notochord has developed into a segmented **vertebral column**, or backbone. The first vertebrates were various kinds of fishes, and paleontologists often call the period from 350 to 400 million years ago the "age of fishes." Fishes are still the most numerous vertebrates.

The simplest fish alive today are the **lampreys**. These animals have no fins and are shaped like large, flexible lancelets, but their skeletons, made of soft **cartilage** instead of bone, show that they are true vertebrates. Larval lampreys are filter-feeders like lancelets. As the animal grows, however, it turns into a predator that attacks other fish. In the adult, the lattice between the pharyngeal slits becomes hardened by cartilage, and the structure serves as a framework for gills. As gill slits they still strain water, but now it is oxygen that is being extracted from the water instead of food.

Lampreys have no jaws – their mouths are round, like suction cups, and they attack their prey with a rasp-like tongue. Because of this, lampreys are considered "jawless fishes." The term *cartilaginous fishes* is reserved for those fish with cartilaginous skeletons that do have jaws, **sharks** and their relatives the skates and rays. In these animals we can see where jaws came from – they are simply the first set of cartilage gill arches in modified form

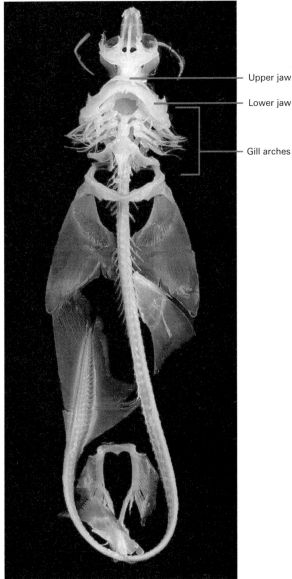

Figure 21.8 *Cartilaginous skeleton of a shark, showing jaws and gill arches*

251

Figure 21.9 *The skeleton of a perch*

(figure 21.8). Note how jaws have arisen independently from key characteristics of different groups: in arthropods the mandibles are a type of segmental appendage, but in chordates they are formed from the slit structure of the pharynx. Jawed fish are good predators, which means they benefit from an active lifestyle. Thus cartilaginous fish have numerous fins to make them powerful swimmers.

Most fish today are bony fish, not cartilaginous fish. They have intricate, rigid skeletons hardened by mineral deposits (figure 21.9). Their skeletons still have the basic segmented character of more primitive chordates, although the segmentation is obscured by bony rays that support the fins. Fish muscles are also quite segmented (figure 21.10), which accounts for the delicate flaky texture of the flesh of cooked fish.

The digestive tract of a fish is subdivided into esophagus, stomach, and intestine, as it is in many animals (figure 21.10). Associated with the digestive system of vertebrates is a large organ called the **liver**. The liver produces many proteins and other blood components as well as enzymes for the digestive system. It is very important in maintaining the animal's body chemistry. Dorsal to the digestive tract of a bony fish is a balloon-like **swim bladder**. It is an outgrowth of the esophagus but has nothing to do with digestion – it makes the fish buoyant so that it can spend less energy swimming. Sharks do not have swim bladders, and they sink when they stop moving.

AMPHIBIANS AND REPTILES

In most bony fishes the swim bladder is separate from the esophagus, even though they arise from the same tissue. In some species, however, they are connected at the pharynx. Lungfishes can take advantage of this connection to use the organs as **lungs**. This means that they can gulp air in through their mouths and hold it inside their bodies to oxygenate their blood without using gills. This same arrangement is found in all vertebrates that live on land. Even in the most modern vertebrates, the access to the lungs is through the pharynx, which is why people can choke if they try to inhale and eat at the same time. In this respect, lungs are not as efficient as the tracheal system of insects, which is separate from the digestive tract.

Some types of fossil fishes had fins that were bulky and muscular. Fossils from 300 to 350 million years ago show that over time in some species these fins became more and more like legs. From fish with legs and lungs came the first **amphibians**. Amphibians are only partially adapted to terrestrial life. For example, their lungs are small and inefficient balloons, so most of their gas exchange occurs through their thin, moist skin, just as it does in earthworms and snails. As a result, amphibian skin is not as waterproof as an insect cuticle. Furthermore, amphibian eggs are highly susceptible to desiccation and must be laid in water. Thus, amphibians are always tied to water for at least part of their lives.

Amphibians and all terrestrial vertebrates are **tetrapods**, meaning that they have four limbs. In no species are there more than four limbs, although some animals have fewer through reduction (e.g., snakes). In many species that lack limbs the skeleton retains small vestigial bones where one would expect to find the missing legs. Moreover, the bones in a tetrapod limb follow a definite pattern: a single bone extends from the body to the first joint (elbow or knee), a pair of bones extends to the next joint (wrist or ankle), and the limb ends in five fingers or toes. Again, this pattern can be modified slightly in some animals – several bones can

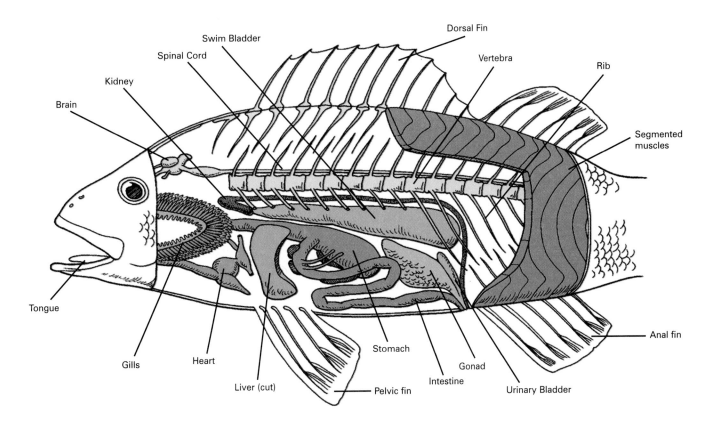

Figure 21.10 *Diagram of the anatomy of a perch*

be fused to look like a single structure (although there are usually traces of this fusion in the final structure) or some digits may be lost. But the basic tetrapod pattern, originating in the earliest amphibians, has persisted for millions of years (figure 21.11).

Reptiles were the first fully terrestrial vertebrates. They appear in the fossil record about 300 million years ago, and they replaced amphibians as the dominant terrestrial vertebrates during the same period that seed plants replaced lycopods as the dominant form of plant life. Reptile skin is toughened with the protein **keratin** to create an effective barrier against water loss. An additional layer of keratin **scales** makes the skin even more waterproof (figure 21.12). This means, however, that no oxygen can be absorbed through their skin, so reptilian lungs must be more efficient than amphibian lungs. In reptiles and their descendants, the lungs have complicated patterns of folds and bumps that create a huge surface for gas exchange. As a result, their structure is more like a sponge than a balloon.

Reptiles also lay **amniotic eggs**. The amnion is one of several membranes inside the egg that seals the embryo within a fluid environment. The combination of dry skin, scales, efficient lungs, and amniotic eggs is so effective in dealing with dry environments that reptiles are especially well-adapted to deserts.

THE DESCENDANTS OF ANCIENT REPTILES

The first reptiles were not tied to water like amphibians, and they were larger than insects; they had little competition and quickly diversified. The main line of reptiles developed into many groups, the most successful being **archosaurs** – crocodiles and alligators, dinosaurs, and birds.

Archosaurs are more active than other reptiles, having higher metabolic rates and body temperatures. Crocodiles use this extra energy for quick predatory movements. They do not have the squat posture of lizards, with their legs sprawling out at the sides. Instead, crocodilian legs are positioned closer to their bodies so that their feet are under their center of gravity. This raises the body into a posture conducive to running. Dinosaur skeletons exhibit an even more erect posture, and these creatures are believed to have been fairly active.

Birds are the most active archosaurs of all. Their metabolic rates are especially high, and they maintain a warm body temperature no matter what the weather. They retain this heat with a thick insulation of **feathers**. Feathers are made of keratin and are actually scales that are modified into a complex, lightweight latticework (figure 21.13). The combination of feathers and strong,

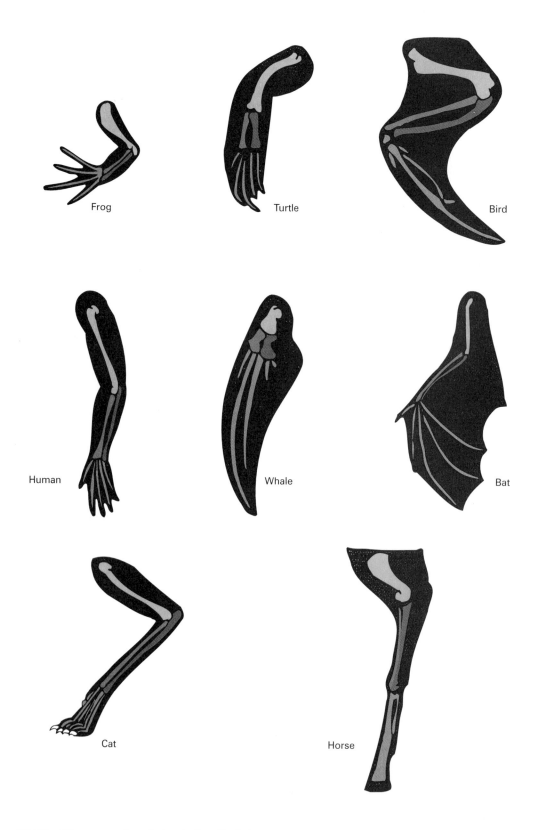

Figure 21.11 *Diagram of vertebrate forelimb bones*

active muscles allowed birds to develop flight, a way of life that has had a major influence in shaping the skeleton and form of modern birds. The laws of aerodynamics have molded bird wings into a flat shape similar to insect wings, but their internal structures are quite different. Bird wings are modified limbs, not extensions of an exoskeleton. Because of the tetrapod limb pattern, birds have only two legs left for walking or running.

Archosaurs exhibit complex patterns of behavior, especially in rearing their young. When turtles or lizards lay their eggs, they abandon them, and the hatchlings must fend for themselves. In contrast, crocodiles and alligators build nests and guard their young in a manner similar to birds. Fossilized dinosaur nests indicate that at least some species of dinosaurs did the same.

Mammals are descended from reptiles that were not archosaurs, and they are structurally somewhat different from other vertebrates. Mammals did develop a high metabolic rate similar to that of birds, and they use **hair** as an insulator to retain heat. Hair is also made of keratin and, while not as closely related to scales as feathers are, it does develop from similar embryological skin structures. Hair is not as good an insulator as feathers, but mammals can occasionally overheat, so mammalian skin is studded with **sweat glands** to cool the skin when necessary. Some sweat glands are modified to produce **milk**, a fluid fortified with protein and sugar so that it is a nutritious food for babies. These glands are called **mammary glands**. In the most reptilian mammals, the monotremes, milk oozes from a large patch of the mother's skin just like sweat, and her babies lap it up from her fur. In most mammals, though, the milk is delivered through specialized **nipples**. Mammals with nipples also have **lips** and muscular faces so that they can nurse (birds and reptiles do not have lips, and their faces are mostly skin and bone). In many mammals, these traits have been further modified to allow for complex forms of communication, both by facial expression and vocalization.

Figure 21.12 *A lizard, showing reptilian scales*

Figure 21.13 *Diagram illustrating the structure of a feather*

EXERCISES WITH ARTHROPODS AND CHORDATES

A. Crustacean structure

1. Study the external structures of a preserved crayfish. Appendages are apparent on every segment. How many segments are in the abdomen? The segments of the cephalothorax have fused, but you can still count how many segments make up this region by counting the pairs of attached appendages; how many are there? (Don't forget to count the mouthparts and antennae!)

2. Starting at the posterior end, carefully remove the appendages from one side of the animal and lay them out on a piece of paper – do not forget the mouth parts or the flattened appendages that look like tail fins. Remember that the gills are part of the walking legs, but do not be surprised if they do not come off with the legs – the attachment is weak.

3. The typical walking leg of a crustacean has five sections. Which of the crayfish appendages retain these five parts? Note that on the cheliped, the "thumb" is actually the last section, and the "hand" is simply the next to last section with an extension off to the side.

4. The cephalothorax of the crayfish is covered by an extension of the exoskeleton called the carapace. After you have removed the appendages from one side, clip away the carapace on the other side to reveal the gills. There should be one feathery gill at the origin of each walking leg. How would this arrangement facilitate respiration?

5. Cut down the dorsal midline of the carapace so you can remove the entire side. Use forceps to remove the muscle that is exposed, taking care not to damage the underlying structures. Cut through the cuticle of the abdomen, removing muscle as you go. When the viscera are exposed, trace the path of the digestive system.

6. In your notes, diagram what you have seen. Be sure to note the body regions and the number of appendages on each.

B. Insect structure

1. Next examine a preserved insect (grasshopper, cockroach, etc.) under a dissecting microscope and identify the three body regions. Can you find the dividing line between the segments of the abdomen? Of the thorax? Hold the insect on its side so you can see the spiracles along the side of its abdomen.

2. Remove one of the back legs and find the five limb parts as you did with the crayfish. Examine the head under the dissecting microscope and find the mouthparts, removing them one by one with forceps in order to get a good look at their structure and arrangement.

3. Now pin the body, dorsal side up, in your dissecting tray, using one pin in the head and another at the end of the abdomen. Stretch the body out slightly as you pin it down. Be sure that the insect is positioned near the edge of the tray so you can view it under the dissecting microscope – it is impossible to see these structures without magnification! Rotate the wings to the side and clip them at the base with scissors to remove them. How do the wings compare to the appendages?

4. Make two lengthwise parallel cuts down the sides of the dorsal surface, starting at the posterior end. Clip the exoskeleton between these cuts at their anterior and posterior ends. You should now be able to pry off the rectangular piece of cuticle. Use a needle to tease loose any tissue that may stick to the rectangle.

5. If you have cut carefully, you may see the dorsal aorta down the midline. The inside of an insect is filled with soft, greasy white tissue. This is the **fat body**, a loose aggregation of cells that store food and produce important biochemicals, similar to a vertebrate liver. The silvery threads throughout the body are the trachea. Use a wash bottle to vigorously rinse as much of the fat body away as possible.

6. Once the water has drained off, you will be able to see the rest of the internal organs. Look for the parts of the digestive system labeled in figure 21.4. As you probe the intestine with a pin you should expose some yellowish threads – these are the Malpighian tubules.

7. Draw what you see as you did in exercise A. Note the colors and textures of the trachea and Malpighian tubules so you know how to tell them apart.

C. Arthropod diversity

1. Survey the other crustaceans on display, comparing them to the crayfish you dissected. Pay careful attention to the microscopic animals. You can identify these tiny creatures as arthropods by their exoskeleton and jointed appendages.

2. Examine the different insects on display. Note how their appendages, especially their mouthparts, are modified for different lifestyles.

D. Deuterostome embryology

1. Examine a prepared slide of sea star development. Identify the blastula and gastrula embryonic stages. See if you can find larvae with complete digestive systems.

2. Next look at the slides of neural tube formation in a frog embryo. Note how the neural tube forms from an outfolding of the embryonic skin.

E. Invertebrate chordates

1. Examine one of the preserved specimens of a lancelet under a dissecting microscope, and compare it to figure 21.7. Do you see evidence of segmentation?

2. Study the dorsal side of the animal. Can you locate the nerve cord and notochord?

F. Vertebrate dissection: a perch

1. Obtain a preserved perch and examine its external structure. Find the flap of skin covering the gills behind the head. Pull this flap back and trim it off with scissors.

2. Carefully cut down through the body wall from the gills to the ventral side of the fish, making sure that you do not damage any internal organs. Cut across the bottom of the fish until you reach the **cloaca**, the combination of anus and urogenital opening. Cut back up the side of the fish, and then forward to the gills again. You should now be able to remove a large section of the body wall and expose the internal organs.

3. Examine the "leafy" gills – the many thin layers help to expose as much blood to oxygen as possible. Identify the parts of the digestive system shown in figure 21.10. Cut into the stomach and look for indications of what the fish might have eaten. Find the heart and trace the large vessels attached to it. Does one lead to the gills?

4. Cut around the sides of the head above the eyes and remove the top of the skull. Locate the brain and examine the lobes under a dissecting microscope. How does this compare to the brains of annelids and arthropods?

G. Vertebrate diversity

People are generally very careful about collecting and killing vertebrate specimens, so there may be strict limits as to what specimens may be available for your study.

1. Examine the skeletons on display and look for segmented characteristics. Find the 1-2-5 pattern of bones in the various vertebrate limbs, and note how fusion can alter the pattern.

2. Study the body coverings of any reptiles and birds on display. Compare the scales and feathers. Do you find any reptilian scales on the birds? Observe how feathers are arranged on the birds in specific tracts.

3. Observe the skins of any mammals on display. Compare the hair to the feathers of the birds. Also compare the facial features to those of the birds and reptiles. What differences do you see?

H. Vertebrate bones

Vertebrate bones are made of dense tissue impregnated with minerals. If the minerals are removed with acid, they become flexible; if bones are baked to dryness, the minerals leave a very hard but brittle structure. Observe these characteristics in the prepared chicken bones on display

IMPORTANT TERMS

abdomen	crop	lamprey	scale
amniotic egg	cuticle	lancelet	shark
amphibian	deuterostome	lips	spiracle
antenna	dorsal nerve cord	liver	stomach
arachnid	echinoderm	lung	sweat gland
archosaur	exoskeleton	Malpighian tubules	swim bladder
arthropod	fat body	mammal	swimmeret
biramous	feather	mammary gland	tail
carapace	filter feeding	mandible	tetrapod
cartilage	gastric ceca	milk	thorax
cephalothorax	gastrula	neural tube	trachea
cheliped	hair	nipple	vertebral column
chordate	head	notochord	vertebrate
cloaca	insect	pharyngeal slits	walking leg
compound eye	keratin	protostome	wing

Laboratory Report The Origins of Land Animals

FACTUAL CONTENT

Define the following terms. Brief definitions are given in the chapter, but you may also wish to consult your lecture notes and text for a better idea of how the terms are used. Pay close attention to those words that look similar but have different meanings – you will want to be able to tell them apart if they appear next to each other on a multiple-choice exam.

abdomen	crop
amniotic egg	cuticle
amphibian	deuterostome
antenna	dorsal nerve cord
arachnid	echinoderm
archosaur	exoskeleton
arthropod	fat body
biramous	feather
carapace	filter feeding
cartilage	gastric ceca
cephalothorax	gastrula
cheliped	hair
chordate	head
cloaca	insect
compound eye	keratin

lamprey	scale
lancelet	shark
lips	spiracle
liver	stomach
lung	sweat gland
Malpighian tubules	swim bladder
mammal	swimmeret
mammary gland	tail
mandible	tetrapod
milk	thorax
neural tube	trachea
nipple	vertebral column
notochord	vertebrate
pharyngeal slits	walking leg
protostome	wing

In addition to knowing what these terms mean, you should be able to identify, on both diagrams and prepared specimens, those that are anatomical structures.

After the common name of each group of animal, write the formal scientific name of the corresponding phylum:

insects, crustaceans, and spiders

lancelets and vertebrates

A. Crustacean structure

1.   Sketch the external appearance of a preserved crayfish:

How many segments are in the abdomen?

How many pairs of appendages are on the cephalothorax?

Which of the crayfish appendages retain all five parts?

Sketch the appearance of a single walking leg, showing the attached gill:

How would this arrangement facilitate respiration?

2. Diagram and label the digestive system of a dissected crayfish:

B. Insect structure

1. Sketch the external appearance of a preserved insect:

Label the three body regions.

How many legs does the insect have?

How many mouthparts does the insect have?

How do the wings compare to the appendages?

2. Diagram and label the digestive system of a dissected insect:

What colors are the trachea and Malpighian tubules? Do they have different textures?

C. Arthropod diversity:

Sketch or describe the various crustaceans on display:

Sketch or describe the various insects on display:

Note how their appendages, especially their mouthparts, are modified for different lifestyles:

D. Deuterostome embryology

1. Sketch the stages of sea star development:

Label the blastula and gastrula embryonic stages. See if you can find larvae with complete digestive systems.

2.   Sketch the stages of neural tube formation in a frog embryo:

E. Invertebrate chordates

Sketch a preserved specimen of a lancelet:

Do you see evidence of segmentation?

Can you locate the nerve cord and notochord?

F. Vertebrate dissection: a perch

Sketch a dissected perch:

Label the parts of the digestive system shown in figure 21.10.

Cut into the stomach and look for indications of what the fish might have eaten:

Indicate the heart on your diagram. Does one of the large vessels lead to the gills?

Sketch the brain and indicate the lobed structure:

How does this compare to the brains of annelids and arthropods?

G. Vertebrate diversity

Examine the skeletons on display – what segmented characteristics do you see?

Study the body coverings of any reptiles and birds on display. Do you find any reptilian scales on the birds? Where?

Observe the skins of any mammals on display. Compare the hair to the feathers of the birds. What differences do you see?

Compare the mammalian facial features to those of the birds and reptiles. What differences do you see?

H. Vertebrate bones

Describe the treated chicken bones on display – how do the different treatments affect the bones?

Systematics

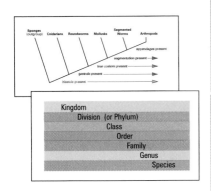

INTRODUCTION

In the last five chapters we looked at several broad categories of organisms. Naturalists have spent centuries studying millions of specimens so that useful generalizations could be made about them, and the observations they have recorded represent an overwhelming amount of data. Since the time of Aristotle, scholars have recognized the need for some kind of biological classification system to help organize all of this information. For most of history, classifiers focused on plants, because they were the mainstay of medicine. Throughout the Middle Ages they produced lavishly illustrated books called herbals that listed and described plants along with their traditional uses. Modern methods of classification are much more complex and can be applied to all five kingdoms of life.

TAXONOMY

During the Renaissance, improved communication within Europe, as well as increased trade between Europe and Asia, Africa, and eventually North and South America, meant that information about many plants from many different areas was coming together. A major problem arose, because all of these plants were called by their common names. For example, many places have a shrub or tree named "ironwood" because it has the hardest wood the people in that area know. But the ironwood tree of one locality may not be the same kind of tree as the ironwood of another locality – they could be completely different plants whose only thing in common was their relatively hard wood. On the other hand, oftentimes a single plant could have several different names if it grew in several different places. These situations created more and more difficulties for the compilers of herbals.

At the end of the sixteenth century, a Swiss botanist named Gaspard Bauhin introduced a systematic way to name plants by giving each plant a Latin binomial, or two-part name. The first part of the **binomial** is the generic name for that type of plant. For example, all oaks are designated by the Latin name *Quercus*. The second part of the binomial indicates what specific type of plant is being referred to – for example, white oaks are *Quercus alba*, red oaks are *Quercus rubra*, and bur oaks are *Quercus macrocarpa*. The generic category (in this case, *Quercus*) is called the **genus**, and the specific type is called a **species**. The binomial therefore consists of a genus name followed by a specific epithet. Binomials are always either underlined or italicized, and the generic name is capitalized. Modern binomials are recognized all over the world, so their use avoids the ambiguities associated with common names.

In the eighteenth century, Carolus Linnaeus developed a new system for classifying plants and animals that employed the use of binomials. Before Linnaeus, most people built their classification schemes from the top down. Aristotle, for example, began with the animal kingdom and looked for ways to divide it up, finally settling on characteristics such as whether the animals had red blood or not. Top-to-bottom approaches tend to incorporate whatever assumptions the classifier begins with, even if the specimens themselves contradict those assumptions. Linnaeus began at the bottom with careful descriptions of species and then looked for similar species to form a genus. He then looked for similar genera to make families, and so on. In this way he constructed a hierarchy of categories which, with but few modifications, is still used today:

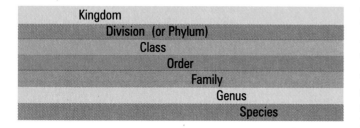

Thus, genera that resemble each other make a family, related families constitute an order, similar orders come together as a class, complimentary classes make up a phylum, and appropriate phyla are included in the same kingdom. With this prescribed methodology based on observations (that is, the descriptions of the individual species), classification became a true science. The science of naming species and studying them to determine where in the classification scheme they belong is called **taxonomy**. Therefore, any particular group in the scheme (e.g., the plant kingdom, the ape family) is called a **taxon**.

In our survey of diversity we looked at the five kingdoms and several divisions in each kingdom. Although we did not mention classes *per se*, we did occasionally focus on some. For example, tapeworms make up Class Cestoda of Phylum Platyhelminthes, and insects constitute Class Insecta of Phylum Arthropoda. Similarly, bony fishes, amphibians, and mammals belong to Classes Osteichthyes, Amphibia, and Mammalia, respectively, of Phylum Vertebrata. We have also used generic names frequently, and these are easily recognized in the text because they are italicized and capitalized. Orders and families are intermediate categories. An example of an order is Carnivora of Class Mammalia. It is made of several families, including the dog family (Canidae), cat family (Felidae), and bear family (Ursidae).

PHYLOGENY

Linnaeus's system was unusual for another reason: it was highly artificial. He did not concern himself with all of the available observations about a particular species. Instead he set up particular rules and focused on the data that the rules called for. When classifying plants he counted flower parts and grouped them accordingly (e.g., plants with two stamens, plants with three stamens, etc.). This made his method easy to use and clear enough to apply to newly discovered species, but most people were interested in looking for taxonomic relationships that were deeper than those proposed by Linnaeus. Accordingly, his methods were soon modified into so-called natural systems of classification – systems that emphasized overall similarities of form rather than arbitrary classification rules. Nevertheless, these new systems still reflected Linnaeus's influence, and they used detailed species descriptions as the basis for building up larger taxa according to the hierarchical plan he devised.

This approach led biologists to the concept of **homologous structures**. Homologous structures are different characteristics whose hidden details exhibit striking similarities. The wing of a bird is homologous to a human arm, because both the wing and the arm contain the same basic tetrapod limb bone arrangement. This similarity persists even though the wing is used for flight and the arm is used for manipulating objects. All tetrapod limbs are homologous because of that structure, no matter what they are used for. Likewise, the dorsal nerve cords of all chordates are homologous because they are all hollow, being derived from the same type of embryological neural tubes. Molecules can be homologous too. The protein in mammal milk has a very similar amino acid sequence to an enzyme found in vertebrate tears. The tear enzyme is a kind of antiseptic that attacks bacteria and prevents them from growing on the moist eyeball. This function is missing in milk protein, but the homology rests on the shared amino acid sequence, not the function.

The opposite of a homologous structure is an **analogous structure**. Analogous structures look superficially alike but are quite different when examined closely. For example, a butterfly wing is flat and membranous, like the wing of a bat. But the butterfly wing contains no bones as the bat wing does – it is an outgrowth of the exoskeleton. The two types of wings are actually different methods for flying molded by the laws of aerodynamics into similar shapes. In practice, it is not always easy to separate homologies from analogies, as we saw when considering the similarities between parapodia and biramous appendages.

While homologous structures obviously indicate some kind of relationship between the species they belong to, no one really understood what that relationship was until Darwin published his theory of evolution. Darwin explained homologies as the result of **divergent evolution**, the process by which the various offspring of a species may slowly adapt to different environments.

If both birds and humans are seen as the modified descendants of a single ancestral tetrapod species, then the similarities in limb structure are simply a reflection of that genetic relationship. On the other hand, analogous structures are explained as the products of **convergent evolution**, the process of adaptation that causes different structures to assume similar forms in order to perform similar functions. This process guarantees that all wings will be flat and broad whether they are made of skin and bones or chitinous cuticle.

Traditionally, taxonomy was studied because of its usefulness, its ability to organize vast amounts of data in a manageable framework. But the theory of evolution offers taxonomy a theoretical basis. Today biologists seek to classify organisms according to how we reconstruct their evolutionary relationships. The evolutionary descent of a taxon is called the **phylogeny** of that taxon, and evolution-based classifications are called phylogenetic systems. Phylogenetic systems often change as new data are uncovered about fossils and homologies, but they are always intended to reflect to the fullest extent possible what is known about the evolutionary history of life. Because of this we no longer think of a species as simply a group of similar creatures, or a specific type of some more generic form. In an evolutionary sense, a species is a population (or group of populations) of individuals capable of interbreeding. The various species of a genus are derived from a single species, but they have now become different enough that they are reproductively isolated and have difficulty interbreeding. The genera of a family are even more distantly related, and so on.

CLADISTICS

Current classification systems retain many pre-Darwinian elements, but they are the focus of much research and debate. The modern study of biodiversity is now called **systematics**, a field that includes both taxonomy (now thought of mainly as the description and naming of species) and phylogenetics. As a historical science, phylogenetics must address phenomena that occurred long before there were any humans to make observations of those phenomena. Therefore phylogenetic models are theories that must be repeatedly tested against the data obtained from specimens. They can also be subjected to logical and mathematical analyses to test for internal consistency. In this way the reliability of any particular phylogenetic reconstruction can be measured and compared to competing theories.

A model of descent is often expressed graphically as a **tree** rooted in an ancestral group and branching out to the various groups descended from that ancestor, as suggested by studies of homologous structures. In the method of **cladistics**, the presence or absence of homologs, and how they may have changed through history, is analyzed with mathematical rigor. First, the traits of interest, designated **characters**, are described in various **character states**. Ideally, one then determines which state of each character is **ancestral** and which are **derived**. Usually this is done by comparing the **in-group**, the group of taxa being studied, to an **out-group**, a taxon related to, but not actually within, the in-group. Whatever character states are in the out-group are designated as ancestral, so that the in-group can be arranged according to changes to the ancestral characters. Each taxon is then given a numerical description for each character state (with the ancestral state, if it is known, being zero), and a **matrix** is constructed showing all of the character states for each taxon.

A cladistic analysis will reveal how ancestral and derived characters are shared among the members of the in-group. **Shared ancestral characters** will be found throughout the group and will not tell us how the ancestral tree branched during history. It is the **shared derived characters** (also called **synapomorphies**) that are the basis of the branching patterns. A tree constructed by this method is called a **cladogram**, and branches of the tree are called **clades**.

We can use our knowledge of invertebrate animals to construct a simple example of the process. We can arrange the various animal phyla based on whether they have certain basic structures. As an out-group we will use sponges, which exhibit none of the listed characters (Table 22.1).

	blastula	gastrula	true coelom	segmentation	cuticle/appendages
sponges	0	0	0	0	0
cnidarians	1	0	0	0	0
roundworms	1	1	0	0	0
mollusks	1	1	1	0	0
annelids	1	1	1	1	0
arthropods	1	1	1	1	1

Table 22.1

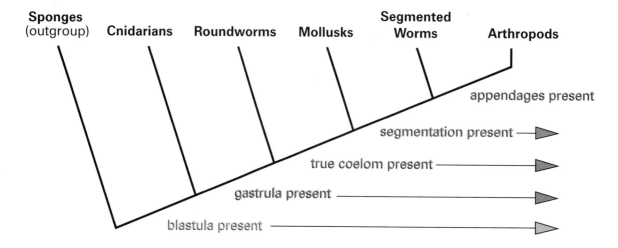

Figure 22.1 *Cladogram based on the character matrix*

This matrix yields a cladogram (figure 22.1) that keeps the phyla together as long as they share derived characters. Segmented worms and arthropods share the most characters on this table, so they end up close together at the top of the tree.

Simple problems like this can be analyzed intuitively. The most interesting situations are complex and require the use of a computer. Computers can evaluate various alternative trees and determine which possible tree is the most **parsimonious** – that is, the pattern that accounts for all of the evolutionary changes with the fewest steps and simplest branching pattern.

EXERCISES WITH SYSTEMATICS

A. Taxonomy of a group of specimens

Combining taxonomy and phylogenetics can be complicated. It is easier to work with classical taxonomic principles first. In this exercise you will be given a set of related specimens. What specimens you use will be determined by what is available, but the basic procedure is the same no matter what you use. Your task will be to group them according to observable similarities and thereby construct a "natural" system of classification. Try to create a hierarchy of sets – small groups of very similar specimens organized into increasingly larger, more generalized groups. You should be able to justify why you put different individuals together at each stage.

1. Start from the bottom up: find two or three very similar specimens for each subgroup. Then bring the subgroups together in increasingly larger groups until all are unified in a single organized "kingdom." You may base your scheme on any characteristics that seem important.

2. Your lab instructor will use several schemes to illustrate how classifications are useful. Expect to see many different approaches. How does your system compare to those made by other students? Is there an obvious "right" or "wrong" system? Are some systems more *useful* than others?

B. Phylogeny of land plants

Before attempting a computer-assisted cladistics analysis, you must know how to analyze character states and complete a proper data matrix. The following exercise is similar to the example we used earlier in building a phylogenetic tree for invertebrates. Use what you learned in chapters 18 and 20 to complete the matrix, and then use the matrix to construct a tree similar to figure 22.1.

1. Complete the following matrix:

	multicellular sporophyte	vascular tissue	microscopic gametophyte	seeds	flowers
Coleochaete	0	0	0	0	0
angiosperms					
conifers					
ferns					
lycopods					
mosses					

2. Search for shared derived characters around which to build groups. Construct a tree to represent these groups.

3. Which group is closer to seed plants in this analysis – ferns or lycopods? What would happen if you only had the first three characters to work with – which groups would be unresolved? What do you think would happen if you had six or seven characters to work with?

C. Phylogeny of a group of organisms

For this exercise, each group will be given a different data set to analyze. You will construct a data matrix and enter it into a computer program called MacClade. MacClade was written by Wayne Maddison and David Maddison and originally distributed as freeware, so it is widely used by biologists (both for teaching and research) even though it is a fairly simple program. Despite its simplicity, MacClade can produce very useful models. The program is designed to let the user build various trees from a data set and compare them in order to see what each tree suggests about the evolution of the characters under study. It does not automatically find the most parsimonious tree as some other programs do, but it does calculate and display a consistency index (C.I.) that reflects how parsimonious each tree is. A consistency index of one indicates maximum parsimony.

1. Examine your data set and construct a data matrix. The matrix will be larger and more complex than what you worked with before.

2. Turn on the computer and launch the MacClade program by double-clicking the program icon.

3. From the program, open one of the scratch files to hold your data matrix. Use the "Save As" command to save the file under your name.

4. Edit the file to show your data matrix. Do not alter the format already in the file or MacClade will not be able to read it. When you are done, save the file to disk again.

5. Use the "Get Tree" command to display your first tree. Start with the default tree generated by the computer. It will not necessarily be the best tree. Look for the C.I. at the left of the screen. As you work, you will want this value to get as close to 1 as you can make it.

6. You will now manipulate the tree by clicking the mouse on portions of the tree and "dragging" them to new locations. Each rearrangement can change the C.I. Mold the tree so as to optimize the C.I.

7. You can make changes blindly, or based on preconceived notions about your organisms. You can also use the traits for help. Click on the "Trace Character" icon (it looks like a miniature tree) to display a trait on the tree. A menu will appear to allow you to select which trait you want to look at.

8. When you have the best tree you can find, copy it into your report or notebook. What does your tree tell you about the evolution of the traits you are looking at?

IMPORTANT TERMS

analogous structure	class	in-group	shared ancestral character
ancestral	convergent evolution	kingdom	shared derived character
binomial	derived	matrix	species
character	divergent evolution	order	synapomorphy
character states	division	out-group	systematics
clade	family	parsimonious	taxon
cladistics	genus	phylogeny	taxonomy
cladogram	homologous structure	phylum	tree

Laboratory Report Systematics

FACTUAL CONTENT

Define the following terms. Brief definitions are given in the chapter, but you may also wish to consult your lecture notes and text for a better idea of how the terms are used. Pay close attention to those words that look similar but have different meanings – you will want to be able to tell them apart if they appear next to each other on a multiple-choice exam.

analogous structure	in-group
ancestral	kingdom
binomial	matrix
character	order
character states	out-group
clade	parsimonious
cladistics	phylogeny
cladogram	phylum
class	shared ancestral character
convergent evolution	shared derived character
derived	species
divergent evolution	synapomorphy
division	systematics
family	taxon
genus	taxonomy
homologous structure	tree

A. Taxonomy of a group of specimens

Diagram your classification scheme:

B. Phylogeny of land plants

Complete the matrix:

	multicellular sporophyte	vascular tissue	microscopic gametophyte	seeds	flowers
Coleochaete	0	0	0	0	0
angiosperms					
conifers					
ferns					
lycopods					
mosses					

Search for shared derived characters around which to build groups. Construct a tree to represent these groups:

Which group is closer to seed plants in this analysis – ferns or lycopods?

What would happen if you only had the first three characters to work with – which groups would be unresolved?

What do you think would happen if you had six or seven characters to work with?

C. Phylogeny of a group of organisms

Examine your data set and construct a data matrix:

Copy the best tree you can produce on the computer:

What does your tree tell you about the evolution of the traits you are looking at?

Behavior Within a Population

23

OBJECTIVES

New Skills

No new skills are required for this exercise.

Factual Content

No new terms are introduced in this exercise.

New Concepts

In doing the exercise in this chapter, you will
1. observe how individual variations of behavior within a population can interact to affect the population as a whole.
2. observe how some behavioral strategies in a population may be more successful than others in varying circumstances.
3. explore methods of population sampling as a way of elucidating past behaviors or trait distributions.

INTRODUCTION

We can study a population as a whole using samples and statistics, but the fact remains that the population remains an aggregation of individuals. Sometimes the different responses of the individuals within a population can interact in complex patterns. Sorting out the details of such interactions can be very difficult, especially if the only observations come at the end of a long process that could not be observed as it unfolded. In such a case the elucidation of the process requires a certain amount of detective work.

The following exercise is a simulation of interactions within a population. The details are deliberately left vague here so that those working on unraveling the "mystery" will be forced to examine the data and deduce what happened from the available facts. Your instructor will supply individual students with more detailed directions concerning their separate roles. Chance will also play a part in how the simulation unfolds.

SIMULATING BEHAVIORAL INTERACTION IN A POPULATION

For this exercise, one or more students will be chosen to play the role of "detective." These students will not participate in the first phase of the exercise, and they should leave the room so that they cannot see what takes place. The rest of the class will be given test tubes of an unknown liquid to use as part of the simulation. Some of the liquids may be fairly acidic or basic, so all tubes should be treated with care. Report any spills to the instructor immediately.

1. You will each be given a tube of clear liquid and a disposable pipet. You will also be given a card describing a strategy for interacting with your classmates. Follow the instructions on your card, but do not tell anyone else what your strategy is. The success of each behavioral strategy will not be evaluated until the end of the exercise.

2. In this simulation, an interaction between individuals is represented by two people exchanging 1 ml of the liquid in their respective test tubes. The various liquids are formulated in such a way that they will be altered by certain mixtures and therefore serve as a record of some types of interactions.

3. Exchange samples of solutions with your classmates as indicated by the instructions on your strategy card. Some people will be directed to have more interactions than others – do not be concerned if your strategy seems different from anyone else's.

4. Continue the interactions until told to stop by your instructor. At that point the instructor will begin collecting data by treating some tubes with an indicator solution. If you have indicator in your tube, do not show the results to anyone but the instructor or a designated "detective."

5. The instructor will bring the "detectives" back into the room and describe what they should be looking for. They will be given bottles of indicator so they can obtain more data for their investigation.

Laboratory Report Behavior within a Population

The students designated as detectives will test the tubes of a small sample of students. Record the results:

What hypotheses might explain this distribution of results?

What are some predictions implied by these hypotheses?

What additional samples are needed to test the hypotheses?

Record the results of the additional tests:

Are any hypotheses eliminated? Can further testing narrow the explanation to a single, well-supported hypothesis?

Appendix 1 Maintaining a Laboratory Notebook

Although this manual contains report forms for recording data, your instructor may want you to keep a more complete record of your work in a notebook.

All professional scientists keep records in notebooks. These notebooks detail what experiments were performed when, what procedures were used, and what results were observed. They are an archive of data. In industrial laboratories, notebooks are the evidence on which patents are built, and they are typically maintained under strict and secure procedures.

Laboratory notebooks usually contain graph-ruled paper to facilitate the making of graphs and tables. They are always bound, not loose-leaf, so that pages will not be lost. Extra notes, instrument printouts, and photos of data can be attached to these pages for safekeeping. A well-kept notebook allows a reader to find data and procedures quickly.

The point of keeping a notebook is to keep an accurate record of events as they occur. It is therefore important that you make your notes directly in your notebook during lab. Do *not* make them on scrap paper to be recopied later. While recopying may produce a neater notebook, it usually produces a less complete notebook, as some scrap notes are invariably lost.

The following directions are for a lab notebook suitable for class use. It is similar to a professional notebook in most respects, but it contains some extra material that is useful if the notebook is collected and graded.

Format
1. The notebook is a *permanent* record. All entries must be in non–water-soluble ink. Mistakes should be scratched out with one or two horizontal lines. Although the notebook needs to be legible, accuracy is more important than neatness.

2. Records should be kept on one side of the paper only – the right (front). The back of the page (left side) may be used as scratch paper for calculations. Scratch calculations may be written in pencil. Records kept on the wrong side of the paper will not usually be graded and can be informal.

3. The first two pages should be left blank at first so a **Table of Contents** can be created as the book is filled in. The Table of Contents should list each chapter or laboratory in order, by title, with the first page it appears on. The Table of Contents should be updated continuously.

4. At the top of every page should be written the *title* of the laboratory recorded on that page, the *date* the notes were recorded, and the *page number* (if the notebook does not have pre-printed page numbers).

5. Each new chapter should begin on a fresh page, even if the previous page is not completely filled.

The following sections describe what material should be recorded for each laboratory session.

Introduction
This is a brief overview of the work to follow. It should focus on the conceptual objectives of the laboratory and indicate what results you may expect.

Materials
This section is a reference for those special materials you would need to obtain if you were to repeat the exercises. Under normal circumstances it may just refer to a page in the lab manual for details. Any changes in materials from the referenced procedure should be noted.

Procedures
All that is necessary here is brief summary of the procedure published in the lab manual – it is unnecessary to recopy the procedure verbatim. Again, you may refer to the relevant pages of the manual for details, noting any changes to the published procedure. Any handouts with special procedures can be attached directly to the page. It is convenient to have this section, and the previous two, done before you come to lab.

Raw data
This section should list any measurements, instrument readings, and qualitative observations. Tables are often appropriate for data. Certain pieces of physical data, such as chromatograms and instrument printouts, should be attached to the notebook. (If your lab partner has such data in another notebook, then note that here.) Sketches and diagrams also belong in this section. *Again, all data should be recorded in the notebook as it is observed* – not recopied from scratch paper later. The report forms included in the manual for each chapter can give you ideas for formatting and recording your data.

Formulas & Calculations
Calculated values and the formulas used to derive them should be recorded next. The actual calculations can be written on the

scratch pages or done on a calculator – the formal record consists of a table of the final calculated values and the data used in the calculations. For example, if an exercise involved the densities of several objects, you could make a table giving the mass, volume, and density of each. The mass and volume would be data, as they are actual measurements, but the density would have to be calculated from these data. Such a table would serve as a record of both data and calculations. Complex calculations may require multiple tables. In this example, if the volumes of the objects were not measured directly but calculated from measurements of height, length, and width, then these measurements might constitute a second table. In some cases you may have to complete the calculations after the lab session.

Graphs

All graphs of data should be drawn in the notebook. An experiment may generate many graphs. Some experiments call for graphs to be made on semilog paper. This can be attached to the notebook as needed. Again, you may have to complete the graphs after the lab session.

Conclusions

This is the section where you address the conceptual objectives. Can you see the trends or generalizations that you expected? The questions in the lab manual should help you draw conclusions – answer them here. Try to build a narrative from your thoughts that will provide an explanation of your observations to the reader.

Appendix 2 Measuring the Concentrations of Solutions

There are several ways of expressing the concentration of a solution, depending on what characteristics of the solution are important for a particular procedure.

Percentage solutions (%)

Percentage solutions are the simplest to make because they require few calculations. An X% solution contains X g of solute per 100 g of solution. Since 100 g of water has a volume of 100 ml, it is common to make percentage solutions based on volume: X% = X g of solute per 100 ml of solution. Remember that a percentage solution always has *less* than exactly 100 g (or 100 ml) of water, since the solute accounts for some of the solution's final mass (or volume). The problem with percentage solutions is that they give us no direct information about how many molecules of solute are in a volume of solution.

Molar solutions (M)

The most commonly used measure of concentration is molarity, or *moles* of solute per liter of solution. With a molar solution, we know precisely how many molecules of solute are dissolved in any volume of solution. For example, a 1 M solution of the sugar glucose has exactly 6.02×10^{23} molecules of glucose in every liter of solution. Again, 1 liter of a molar solution contains less than 1 liter of water because the solute also takes up space.

Normal solutions (N)

Acids and bases are commonly measured in normal solutions, which record the moles of H^+ or OH^- ions present. These ions determine the actual strength of the acid or base. For an acid like hydrochloric acid (HCl, having only one H^+ ion per molecule), normality is essentially the same as molarity. However, sulfuric acid (H_2SO_4) releases two H^+ ions from each molecule. Thus, a 1 M solution of H_2SO_4 is actually 2 normal. One liter of a 1 N acid (whether it is 1 M HCl or 0.5 M H_2SO_4) will always neutralize 1 liter of a 1 N base (whether it is 1 M NaOH or 0.5 M $Ca(OH)_2$).

Dilutions

Diluting a single solution can simplify experiments that call for solutions of varying strength. Suppose that you wanted to see what a solution of NaCl does to some cells. Since the effects vary for different concentrations of NaCl, you would need to test a range of NaCl solutions, say 1 M, 0.1 M, and 0.01 M. You could weigh out NaCl three separate times to make these three solutions, but it would be easier to just make a 1 M solution and then dilute it: Take 1 ml of 1 M NaCl and add 9 ml of water. This solution is one-tenth as strong as the first, or 0.1 M. Then you could take 1 ml of this solution and add 9 ml of water to make a second dilution, 0.01 M. A series of dilutions like this is called a *serial dilution*. By using serial dilutions, you can create incredibly dilute solutions in the nanomolar or even picomolar range. It would be impossible to accurately weigh out such small amounts of material.